Teachers' Guide to

A Primer in
Data Reduction

An Introductory Statistics Textbook

A. S. C. EHRENBERG
London Business School

1807 1982

175 YEARS OF PUBLISHING

JOHN WILEY & SONS

Chichester · Waterford RTC · Toronto · Singapore

Copyright © 1982 by John Wiley & Sons Ltd.
Reprinted December 1994
All rights reserved.

British Library Cataloguing in Publication Data:

Ehrenberg, A. S. C.
  A Teachers' guide to A primer in data reduction
  1. Statistics
  I. Title   II. Ehrenberg, A. S. C.   Primer in data
  reduction
  001.4'22   HA29

ISBN 0 471 90086 9

Printed in Great Britain by RPM Reprographics Ltd., Chichester.

Teachers' Guide to
A Primer in
Data Reduction

Contents

v

Preface

The aim of this guide is to provide background when using the Primer to teach an introductory or beginning course in statistics. For example, it

- shows how one can talk the class through an illustration (e.g. "The mean as a mental focus" in Chapter 1);

- explains why frequency distributions are treated <u>after</u> means, medians and modes in the Primer, or why geometric means are not treated at all;

- discusses some issues more fully (like when weighted averages do or do not matter);

- gives further numerical illustrations;

- notes where one may want to introduce numerical illustrations from the students' main subject-areas;

- outlines mathematical proofs which one can give to certain classes (e.g. for the short-cut formula of the variance);

- discusses alternative approaches to certain problems (e.g. the treatment of probabilities);

- notes problems which students tend to raise.

As teachers, most of us find it useful at times to think about why we are teaching what we are teaching. Here the Guide tries to share my personal experience of many years' practical statistical analysis, as well as the experience other teachers and I have gained with the material in the Primer.

The Primer

As its title implies, the <u>Primer in Data Reduction</u> differs from other introductory texts by putting more emphasis on data.

Teaching points in the text are often introduced through a brief numerical illustration. Like the rest of us, students find it easier if a specific example comes first, followed by the general concept or formula. It also gets them familiar with data.

Students have liked the Primer when other teachers used it. Students say they found most of it easy and clear. They liked the book's brevity, and that the treatment is non-mathematical but not "talking down". They also appreciated that the book tries to say what works and what doesn't.

The Primer covers the basic syllabus of an introductory course. It does so in a fairly standard sequence, as the list of chapter headings shows. But the precise order of topics is fairly easy to change (e.g. to take Probability in class before Theoretical Distributions, or Relationships before Sampling).

Objectives of the Primer

The two main aims are to give students

(i)   basic numeracy (e.g. the ability to summarize simple data sets by means of averages, etc.) and

(ii)   some understanding of the main procedures and concepts of statistical analysis.

The latter also requires practice ("learning through doing"). But I would not realistically expect students to become proficient in applying most of the techniques after only one introductory course. In any case, most students will not have to use the techniques themselves.

Students must be able to appreciate other people's uses of regression, or tests of significance, or whatever. But they are not expected to make new discoveries (which is what most of the classical techniques are for) after only a single course in statistics. If students need to apply a technique later in some practical work or project, that will be the time for them really to learn it – with motivation. Hence it is important for them to have some previous acquaintance with the techniques and to become familiar with a basic text, so they can return to it for later reference.

In teaching an introductory course I assume that we aim to concentrate on statistical material that our students are likely to come across in their main subject-areas or subsequent lives. There seems little point in covering topics which will receive no reinforcement through subsequent use, unless formal course syllabuses or examinations require it.

The Length of the Course

The Primer is most suitable for courses lasting one semester or two quarters or terms.

If time presses or the class has to be taken relatively slowly, it is easy to leave out Parts Five and Six (possibly with instructions to students to read them on their own). One can also leave out all or most of Chapters 6 (Probability Models), 8 (Sampling Distributions), 13 (Many Sets of Data) and 14 (Multi-variate Methods).

In addition, class discussion of the later sections of most chapters can be curtailed or omitted, especially for Chapters 5 (Theoretical Frequency Distributions), 10 (Tests of Significance) and 12 (Regression). Students can then be "warned-off" these sections.

If the course spreads over two semesters (i.e. a whole year), there should be time to introduce some additional and more specialised topics, especially towards the end.

Additional Materials

As stressed in the Preface to the main text, some teachers will need to cover additional topics relevant to the main subject-areas of the class (e.g. index-numbers for economists). For such a purpose it is usually easy to refer students to a specific chapter in some other text or to a journal article or two, or to prepare one's own supplementary notes.

The illustrations and exercises in the Primer are from a variety of subjects but are simple and realistic and have been found to interest a wide range of students. The end-of-chapter exercises are given to provide practice, with the answers given at the back of the book.

Additional illustrations can often be useful in class. These should generally be from the students' own subject-areas, to increase their relevance. Points in the Primer where this might particularly apply are noted in this guide.

Exercises for students to do themselves should, I believe, also be from the students' subject-areas, or at least be couched in the relevant language. (Some practice exercises can be fairly easily constructed by adapting the numbers from the exercises in the book and re-casting these in the relevant subject-area language, i.e. in the classical "Three men digging a trench for two days...." manner.)

Computers

The use of computers is naturally referred to at times in the Primer, but no strong link with computing has been written into the text. In my experience, teaching practice in this respect still varies very widely –

from no computing for beginning classes in statistics to a separate course of instruction or a fully integrated treatment.

My own view is that most students need to acquire familiarity with micros or the like, and will increasingly already have some. This is quite apart from their use in formal statistics. But for the latter – at least as treated in an introductory course – it is probably best if students first acquire some pencil and paper familiarity with the calculations, using small numerical examples. That is how the Primer has been structured; computing exercises can then be added.

Acknowledgements

I am indebted to Helen Bloom Lewis for useful comments on the draft of this Guide, and to Myra Davies for typing the final camera-ready copy (and also the tables as reproduced in the Primer). I am also grateful to Ian Shelley of John Wiley & Sons for his helpful support.

Comments from Teachers

I would be very pleased to receive comments and criticisms about the Primer or this Guide. Although the text has been through many drafts and classroom try-outs, I know there is still room for improvement.

London Business School
Sussex Place
London NW1 4SA

A.S.C. Ehrenberg
June 1982

Part One: Statistical Data

In my experience students find no problem in dealing with summary measures in Part One before the formal discussion of frequency distributions in Part Two.   Summary measures seem more practical and less frightening as a starting-point.

Students with a mathematical background may think means, medians and modes a bit trivial, but they are usually not accustomed to dealing with data, and in particular not with skew data.

CHAPTER 1 : AVERAGES

The main lesson throughout the Primer is that if we are to understand and interpret observed data, we need to boil it down to just a few summary figures which we can actually cope with mentally.   In practice, nothing is more effective than that.

1.1   The Arithmetic Mean

Averages are easily the most important tool in statistics (ultimately even standard deviations, regression equations, and statistical expectations are all just averages).

Problems arise with averages for skew or very skew distributions. It is important to give students a feel for some of the more extreme cases, with realistic examples.   (In the text, the numerical illustrations are all made to give an average of 5.4, for ease of exposition.)

1.2   The Mean as a Focus

The role of the mean as a visual or mental focus is often very important in getting a feel of data, but is not usually discussed in statistical texts.

I find it good to "talk the class through a case".   For example, putting the first row in Table 1.6 (i.e. without an average) on the blackboard or on a transparency

$$6, 3, 7, 5, 6, 4, 4, 6, 7, 3, 5, 9, 6, 4, 2, 7, 5, 5, 8, 6,$$

I would first just look at the numbers (with the class) and say that I do not really know which figure to compare with which.   I would act this out by (i) reading the numbers out aloud in front of the class, maybe pointing at each number in turn, i.e. 6, 3, 7, 5, 6, 4, 4, 6, 7, (ii) stop reading at about this point, say that I am looking back again at the earlier numbers to remind myself what they were like ("Yes, there was a 7 earlier .. "), and (iii) lose my place ("where had I got to?") and get all confused.

In contrast, with the mean of 5.4 in mind as a single comparison figure or focus - or better rounded to "about 5" - I can compare each number in turn with just that single value: the numbers 6, 3, 7, 5 etc are successively above 5, then below, above, the same, above, below, below, above, and so on.   I can see that about half the figures are above and half below the mean, and that most are just one or two units away on either side of 5.*

The examples in Tables 1.6 and 1.7 allow us to show that the mean is still a good focus in this way even when the data are highly skew (or U-shaped).   The averages are not typical but still useful: we can for example see any skewness of the data more easily.

Table Lay-Out

In Tables 1.6, 1.7 and 1.7a in the Primer it would have been easier to read across each row if there had been a small gap between the second and the third rows in each table.   The eye could then move much more readily along each row.   You may want to make this point in the class (noting that I got it wrong in the Primer - it is very effective to criticise the course text occasionally!)

To illustrate, you can make a transparency or Xerox of the two versions of Table 1.7 on the facing page.   (Table lay-out is discussed more generally in Chapter 16.)

1.3   Comparing Different Sets of Data

Faced with a single small set of 20 readings as in the illustrations, some students wonder why one needs to work out a summary like a mean at all.   They tend to say that it is pretty obvious what the figures are like just by looking at them.   And the students are right.

Summaries become important with larger sets of data and/or when comparing and interpreting more than one set of data (in practice often many).   It may help here to use illustrations of a larger data-set, or several sets, from the students' own subject-areas.

---

* There is a misprint in the first line of page 8 of the Primer.   The figure in
" (the mean of) each set is about 5, we can ..." should be 5, not 5.5.

TABLE 1.7   The Four Sets of Readings, Ordered by Size

2, 3, 3, 4, 4, 4, 5, 5, 5, 6, 6, 6, 6, 6, 7, 7, 7, 8, 9   Mean: 5.4

2, 3, 3, 4, 4, 4, 5, 5, 5, 6, 6, 6, 6, 7, 10, 15   Mean: 5.4

0, 0, 0, 0, 0, 0, 1, 1, 1, 1, 2, 2, 2, 3, 3, 15, 76   Mean: 5.4

0, 0, 0, 0, 0, 0, 1, 1, 2, 10, 11, 11, 11, 12, 12, 12, 12, 12   Mean: 5.4

TABLE 1.7   The Original Table, with a Gap

2, 3, 3, 4, 4, 4, 5, 5, 5, 6, 6, 6, 6, 6, 7, 7, 7, 8, 9   Mean: 5.4

2, 3, 3, 4, 4, 4, 5, 5, 5, 5, 6, 6, 6, 8, 10, 15   Mean: 5.4

0, 0, 0, 0, 0, 0, 1, 1, 1, 1, 2, 2, 2, 3, 3, 15, 76   Mean: 5.4

0, 0, 0, 0, 0, 0, 1, 1, 2, 10, 11, 11, 11, 12, 12, 12, 12, 12   Mean: 5.4

1.4   Medians and Modes

Seeing where the median or the mode are gives us a good feel for the shape of a distribution.   But otherwise neither measure is widely used, e.g. for numerical calculations.   That is a fact rather than a value judgement.   Thus in most statistical texts (including the Primer), the median and mode are virtually never referred to again once they have been defined.

One reason is that for skew data like an income distribution, the median is <u>also</u> not very typical or representative, i.e. it is not much better than the mean.   (The fun that is often had at the expense of the mean usually comes from misunderstanding it: to say that the average family has 1.4 children does not claim that this is <u>typical</u>.)

It is often said that the mean is more precise than the median or mode in that it takes account of the numerical values of all the different readings.   But this is not very important unless n is very small.

1.5   Outliers

Outliers are a nuisance.   But they are less of a problem – less judgement is needed – when dealing with data which are of a familiar form.

For the illness data in Table 1.14 for instance, what did previous data look like?   Is it at all common to have 1 in 20 readings well over 50? Discussing an example or two of data from the students' own subject-area would be good here.

Students sometimes think that leaving an outlier out may be cheating. But reporting it separately actually highlights it.

1.6   Weighted Averages

Students will gain confidence in dealing with means if they have practice at working out some weighted averages (using examples relevant to the class's subject-areas).

People's initial inclination is usually for precision (i.e. for a weighted average).   But often the effect of weighting is small, and un-weighted averages are simpler to calculate (especially if the weights are not known, as is often the case with reported data !)

There are three main situations to discuss (possibly getting the class to work out the means as you go along):

1. If the $\bar{x}$'s are similar, weighting makes relatively little numerical difference. An example for three countries A, B and C, say (with the weights adding to 50) is:

	Countries			Means	
	A	B	C	Weighted	Unweighted
n :	5	15	30		
$\bar{x}$:	28	30	32	31.0	30.0

2. Similarly, if the weights (n) are fairly similar but the means are very different, weighting still makes little difference (compared with $\bar{x}$'s varying from 5 to 55):

	Countries			Means	
	A	B	C	Weighted	Unweighted
n :	20	15	15		
$\bar{x}$:	5	30	55	27.5	30.0

3. If both the weights (n) and the means differ (e.g. C is a very big country in both senses), then the weighted and unweighted means will differ quite a lot :

	Countries			Means	
	A	B	C	Weighted	Unweighted
n :	5	15	30		
$\bar{x}$:	5	30	55	42.5	30.0

But neither the weighted nor the unweighted mean is then descriptively typical or representative. In such cases it would usually be better to describe the ungrouped data (i.e. the separate countries), just as we often have to describe a very skew distribution in some detail. In particular, the weighted average of 42.5 here mainly reflects the mean of the big country C (dragged down a bit by the two smaller countries with their smaller means).

1.7 Algebraic Notation

Experience shows that students who have little mathematics can happily learn a few common symbols like n, x and $\bar{x}$. But they find algebraic operations difficult, especially at this stage in the course. They may think it is their fault and switch off.

Σ is more complex than n or x because it is an operator. To explain that Σ really stands for $\sum_{i=1}^{n} x_i$ probably makes it worse for the students. And even saying that this in turn stands for $(x_1 + x_2 + \ldots + x_i + \ldots x_n)$ introduces new conventions (like x_i and x_n). Keeping to Sum(x) is conceptually simpler for the newcomer.

It is all rather as if we lectured not only in an unfamiliar language (mathematics) with a strange grammar (the algebra), but also used a special dialect (each teacher's own choice of symbols) and special syntax (like that x_i, x_j and x can all stand for exactly the same thing — but that x_i and x_n cannot, except "of course" when i = n !)

It helps students who are unsure of their mathematical backgrounds if we mention that we, the teachers, usually find other mathematicians' mathematics difficult too.

1.8   Discussion

If there is time, it is good to illustrate the descriptive use of means (and medians and modes) with some larger data sets from the students' own subject-areas.

The crucial idea is that the mean (plus some idea of the nature and size of the scatter of readings) should by itself conjure up a fair idea what the data were like, without anyone still having to see the detailed data.

Thus a mean of 5.4 should usually tell us that the readings are mostly round about 5, and a mean of 45 that the readings are mostly round about 45.

CHAPTER 2 : SCATTER

In class it might be a good idea to introduce the range briefly at the start of this chapter, rather than leaving it all until §2.6 as in the main text. It is a very natural measure.

The range for the example of §2.1 in the Primer is 10. That gives us quite a good feel for the data - without seeing the individual readings - compared with readings which have a range of 2 or one of 50.

A fuller description would give the two end-points separately, i.e. that the readings range from 1 to 11. That tells us more. But it involves giving two values, which is clumsy when comparing different sets of data (e.g. 1 to 11; 15 to 24; 32 to 49; 46 to 54; etc, compared with single-value ranges of 10, 9, 17, 8, etc).

In practice we mostly use two values, but two quite different ones - the mean and a single measure of scatter. With roughly symmetrical data this describes the data well. Thus in the numerical example of §2.1 the mean is 5 and the range is 10. The two end-points must therefore be about 0 and 10, which is almost right (1 and 11 actually). We can then compare different sets of data firstly by their means and then by their scatter. Skew data are, as always, more complex to describe.

But while the range is a common-sense measure to look at (and does give a good feel for the data), it has the major drawbacks discussed in §2.6, e.g. for large sets of data and for comparing sets of data with different n.

2.1 Deviations from the Mean

Summarising the scatter of readings by their deviations from the mean is very common in statistics (especially with symmetrical data). First we give the mean, then the scatter about it, calculated as a measure which does not depend on the number of readings n (i.e. as an average).

To explain to students that the deviations $(x - \bar{x})$ always add up to zero, one can again use the numerical example in §2.1, i.e. 11, 2, 5, 1, 6. The sum of the five deviations from $\bar{x} = 5$ is

$$(11 - 5) + (2 - 5) + (5 - 5) + (1 - 5) + (6 - 5).$$

Rearranged this reads

$$(11 + 2 + 5 + 1 + 6) - (5 + 5 + 5 + 5 + 5) = 25 - 5 \times 5 = 0,$$

i.e. the sum of the n readings minus n times the mean, which always equals zero.

Showing this algebraically, it reads

$$\text{Sum}(x - \bar{x}) = (x_1 - \bar{x}) + (x_2 - \bar{x}) + \dots + (x_n - \bar{x})$$

$$= (x_1 + x_2 + \dots + x_n) - (\bar{x} + \bar{x} + \bar{x} + \dots + \bar{x}) = \text{Sum}(x) - n\bar{x}, = 0.$$

This is zero since $\bar{x}$ is defined as $\text{Sum}(x)/n$ and hence $n\bar{x} = \text{Sum}(x)$.

This sort of simple "explanatory proof" is worth doing in class if there is time and about half the class are able to take it in.

2.2   The Mean Deviation

The mean deviation is a common-sense measure for seeing how big the deviations from the mean are. In the example, the deviations 6, 3, 0, 4, 1 are clearly on average about 3 units (i.e. 14/5).

It is important not to be too dismissive about the md (as some texts are). Students find that difficult. The md is conceptually straight-forward and is widely used in practice, especially if the deviations are already worked out. It is also a good introduction to the sd. Two examples are:

Example A: The mean heights in inches of eight different groups of 12-year-old children are:

56, 56, 58, 59, 56, 58, 62, 61. Mean: 58.

(These figures come from the last column in the body of Table 3.1, but are written as a row here.) It is easy to see by eye that the values differ from the mean something like two inches (2, 2, 0, 1, 2, 0, 4, 3).

This is still easier to see if the individual deviations have already been written out (as in Table 3.4), i.e. as

-2, -2, 0, 1, -2, 0, 4, 3.

Adding these up mentally (ignoring the signs), again gives an md of roughly 2, or more precisely 14/8 = 1.8.

<u>Example B</u>: In assessing the fit of an equation, one can draw a rough graph
on the board or overhead projector for two variables x and y (e.g.
y = 10 + 3x for the numbers of large and small grocery stores in n = 5
districts in Table 12.1.) Having drawn it, one can say to the class that
the (vertical) deviations work out as

4,   -1,   2,   -5,   0, or whatever.

To summarise these, the average or md is 12/5 = 2.4, or "about 2".
That's straightforward and what most of us do in practice.

<u>What Sort of Average is the md?</u>

If one draws something like a Normal distribution on the blackboard,
one can point out that most of the deviations will be near-zero (because that
is where the hump is).   Lots more will be fairly close to zero on either
side of the mean, and only relatively few will be far out.

In fact, about 60% of the deviations usually lie within ± 1 md from
the mean.   The rest lie outside these limits, but mostly only <u>just</u> outside,
i.e. within about ± 2 md's.   Only a few readings lie further away than that,
and even these are mostly not <u>much</u> further.   (For the Normal distribution -
Ch.5 - the three percentages are of course about 58%, 90%, and 10%.)

Even with skew data most of the readings still lie within ± 1 md of
the mean, and nearly all within ± 2 md's, <u>but in an unsymmetrical way</u>.
(One can show this graphically.)   With <u>very</u> skew data, a few readings will
lie way beyond 2 md on one side of the mean.

<u>Drawbacks of the md</u>

Now one can give the bad news: The md is <u>very</u> tedious to calculate
for large data sets, for figures of more than one digit, and when the
deviations are not already written out.

The md is also very difficult to use mathematically in further calculations (rather like the median).   If asked by students to expand on these
difficulties one can briefly look ahead and say that there is no short-cut
formula for the md as exists for calculating the standard deviation (as will
be seen very shortly in §2.4).   One can also briefly mention that md's are
pretty hopeless for (i) the analysis of variance (touched on in Chapter 3),
(ii) the theory of statistical sampling (e.g. the standard error formula in
Chapter 8), and (iii) regression theory (in Chapter 12).

2.3   The Variance and the Standard Deviation

The standard deviation – easier to compute from raw data and much easier to use in theoretical work – is also a way of measuring the average size of the deviations.

The sd and md are rather like weighing people with or without clothes on,   the sd always being a bit larger than the md – by about 25% for symmetrical humped-back distributions.   About two-thirds of the readings lie less than ± 1 sd from the mean (instead of about 60% for ± 1 md), and the other third lie mostly within ± 2 sd's.   (This is useful to say here as a fact of life, to be formalised in Chapter 5 with the 68% and 95% figures for the Normal distribution.)

The variance is just an intermediate step in these calculations – very useful in some theoretical work, but of little use descriptively.

2.4   A Short-Cut Formula

Exercise 2.1 rubs students' noses in the tedium of calculating the sd even with a small n.   The short-cut formula is therefore a great boon – it eliminates the need to calculate and write down the n deviations $(x-\bar{x})$. (No such short-cut is mathematically possible with the md ! !)

For the more mathematical members of the class one may want to outline the derivation of the short-cut formula.   The aim is to express $\text{Sum}(x-\bar{x})^2/n$ as $\{\text{Sum}(x^2) - n(\bar{x})^2\}/n$.   Ignoring the n's we can write

$$\text{Sum}(x-\bar{x})^2 = \text{Sum}(x-\bar{x})(x-\bar{x})$$

$$= \text{Sum}(x^2 - 2x\bar{x} + \bar{x}^2)$$

$$= \text{Sum}(x^2) - 2\bar{x}\,\text{Sum}(x) + n\bar{x}^2,$$

(taking care to explain the derivation of the last two terms explicitly),

$$= \text{Sum}(x^2) - 2\bar{x}.n\bar{x} + n\bar{x}^2$$

$$= \text{Sum}(x^2)^2 - n\bar{x}^2, \text{ as required}.$$

Of the several ways of writing the correction for the mean, $n\bar{x}^2$ seems conceptually simplest, but the $(\Sigma\bar{x})^2/n$ version suffers less from rounding errors.

2.5   Other Properties of the Standard Deviation

<u>The Divisor (n-1)</u>:   Personally I regard this as a big but unimportant
nuisance at this stage – a slight adjustment to the variance and the
standard deviation as straight averages.   It makes virtually no difference
numerically or conceptually here.   However, the formulae are generally
used with (n-1) and so students had better learn to recognise them that way.

(One can say that there are good reasons for using (n-1) that will
become clearer later, e.g. in §8.5 and §9.1.   I believe it is far too early
to try to discuss technicalities like degrees of freedom or lack of bias now,
let alone Anova, contingency tables, multiple regression.   Students appear
willing to accept a "black box" argument which (i) does not really matter
and (ii) will become clearer later.   In my experience some teachers make
much more fuss about (n-1) than students ever do.)

<u>The Coefficient of Variation</u>:   Whether to use the sd or CV is best treated
pragmatically: Which is simpler?   This depends on how the sd varies with
the mean in different sets of data.   It may also be worth pointing out that
the CV makes little sense for variables which can take negative values.

2.6   The Range

As mentioned earlier, the range is a very common-sense measure,
but not of much practical use in detailed analysis.   Not being an average,
it depends on the sample size.   To illustrate this, one can take the ages
of the students in the first row of the class and note how adding students
from the next row must either increase the age-range or leave it as it is,
but can never make it smaller.

With symmetrical humped-backed distributions and n up to 50 or
100, the range is roughly 5 or 6 times the sd.

2.7   Discussion

To illustrate the practical importance of knowing about scatter, one
can refer to a university admissions test where the average score of those
accepted is 550.   This (of course) does not mean that all students have to
get 550 or more to be accepted.   If the distribution is roughly symmetrical,
about half of those accepted got less than the average.

We therefore want to know the size of the scatter: do most of the
scores lie between 500 and 600 (mean 550), or is the effective range for
most of the readings much wider?   Is the distribution hump-backed?

Roughly where on the scale is the cut-off point for acceptances? (And what about the scores of those applicants who did <u>not</u> get accepted?)

More generally, it may now be worth telling the class that so far we have concentrated on two of the three aspects of observed data which usually need to be described: average size, scatter, and distribution shape. The latter has often been mentioned, but will be discussed more formally in Part Two.

In practice one mainly compares different data sets <u>of the same kind</u> (e.g. the heights of different groups of people). The distribution shape and the scatter of the different sets of readings are then often similar. Hence the data-sets differ only in terms of their means. That is why averages are so important. (This leads into Chapter 3, where all the readings in a table are averages.)

CHAPTER 3 : STRUCTURED TABLES

The discussion of structured tables is unusual at this point in introductory statistical texts. But it involves averages and notions of scatter and thus provides a useful recap.

This chapter tends to be popular with students: they usually see statistical data in the form of tables, not as frequency distributions, so the material seems more relevant to them.

The kind of table discussed here is one where there is just a single variable (like people's heights) but many different sets of data. Each reading in the table is the average for one such data set.

3.1   Row and Column Averages

Many tables are given without marginal averages, but they are almost invariably helpful. They provide a visual or mental focus to help us to see the patterns better and sometimes provide useful summaries.

3.2   The Role of Averages

This section has been deliberately structured to reinforce the discussion of the role of averages for frequency distributions in Chapter 1.

Row and column averages (and averages of sub-sets if there are "interaction" effects) not only help one to see the patterns and exceptions in the data, but also to summarise the data. Instead of reporting the full table, averages (and indications of scatter) should often do.

3.3   Deviations From the Mean

Deviations can highlight sub-patterns in the data. Students find this useful.

One tends to summarise the variation in a row or a column by working out mean deviations. This is quick and easy to do here, since the individual deviations have already been calculated.

3.4   Weighted Means and Totals

The number of readings of each cell of a table is seldom reported
(nor the scatter of the readings).   This is all right as long as none of the
conclusions to be drawn from the table would be affected by these features,
e.g. no outliers and very small n's.

As a result one usually has to work out <u>unweighted</u> row and column
averages, or unweighted averages of sub-sets of the data.   But in most
cases this does not matter (as discussed in Section 3.4 of the Primer and
§ 1.6 in this Guide).

3.5   Discussion

There is emphasis these days on "looking at the data", rather than
on just applying statistical techniques.   The discussion in this chapter
illustrates this.   Other aspects are discussed in Part Five.

(Looking at row and column averages is like looking at the main
effects in an Analysis of Variance, which really is an analysis of <u>means.</u>
Technically Anova is of course mainly concerned with tests of significance,
in cases where the data are based on small random samples.   Few
students come across applications of the Analysis of Variance, but many
will see tables of means.)

Part Two: Frequency Distributions

Students have already come across frequency distributions in Chapters 1 and 2. Part Two now treats them more systematically. Chapter 4 deals with the numerical aspects of observed distributions and Chapter 5 with theoretical distributions like the Normal, Poisson, etc. These are treated at this stage as descriptive formulae to model observed data, without any reference to probabilities.

The probabilistic formulations are only introduced in Chapter 6. This sequence is unusual. It is adopted because it seems easier for students.   In Chapter 5 they have to take on board the general concept of theoretical distributions, plus specific properties and formulae, plus the idea of modelling observed data.   This seems quite enough for one chapter, without the rather complex concepts of probability.

I think there is little harm in dealing with probabilities only in Chapter 6.   It is only a week or two later.   (Newcomers to theoretical frequency distributions will not miss the absence of a probabilistic treatment in Chapter 5, since they do not know what they are missing.) But teachers who do not find the sequence of Chapters 5 and 6 suitable can introduce probabilities earlier.   It does not matter too much if the sequence of topics in the classroom is momentarily out of step with the sequence of chapters in the book.

A further reason for treating theoretical distributions as ad hoc descriptive formulae in Chapter 5 is to distinguish this use from their use in the context of random sampling.   This is confused in a number of texts, particularly for the Binomial distribution.

The emphasis in Chapters 5 and 6 on additional theoretical distributions like the BBD and NBD is unusual in introductory texts.   The aim is to make the introductory student aware that a whole variety of possible models exist.   But this can be condensed, or cut in class, if time presses.

CHAPTER 4 : OBSERVED DISTRIBUTIONS

This chapter deals with observed distributions, including grouped data, percentiles, histograms, etc, as may be seen from the section headings listed on page 42 of the main text.

The main distribution shapes, recapitulated in Figure 4.1, have already been introduced in previous chapters. It may be worth adding that more complex shapes can also occur, like multi-modal ones which often arise from mixtures of distributions. (An example would be the income distribution in a rural economy, with a mode for farm labourers and a higher one for landowners.)

4.1 Counting the Frequencies

It is useful here to give students practice in working out means and standard deviations of frequency distributions expressed in terms of x and f.

Teachers familiar with Tukey's stem-and-leaf procedure for tabulating a frequency distribution may also wish to introduce this. But students on an introductory statistical course do not tabulate frequency distributions often and may not need to learn such a technique.

4.2 Relative Frequencies

It is worth lingering for a few moments over relative frequencies. They are a rather brilliant (i.e. simple) invention to eliminate the effect of different n's and thus make comparisons much easier. It is odd how many statisticians in their own work report frequency distributions as absolute numbers.

Students can get confused here over proportions and percentages, if these are not contrasted explicitly.

4.3 Grouped Data

Many students would like a categorical rule for the number of grouping intervals they should use. It is a good time to stress that there often are no hard and fast "right" rules in handling data, i.e. that some judgement is needed.

To guide that judgement, we can tell them that having too many grouping intervals (like 10 or more) nullifies the effects of grouping.   The purpose is to simplify the data by not having to take in too much detail and to smooth out quirks in the data which one is not able to interpret anyway. But if some important details are being lost, then one is using too few grouping intervals.   We can say that there is nothing wrong in being pragmatic between these two extremes, since the whole purpose of grouping is convenience.

Some degree of precision is lost in the process of grouping. Exercise 4.2 gives a numerical illustration.   To enlarge on this it is best to use a numerical example from the students' own subject-areas.

4.4   Percentiles

It is worth while for students to become really familiar with percentiles.   They are the common language for describing frequency distributions (e.g. the percentage of readings lying between two values, or the cut-off points in significance tests).

With cumulative frequencies, it is worth showing that when we leave out certain of the x-values (as in the example at the end of § 4.4) it is equivalent to grouping ordinary frequencies (e.g. 0 to 3, 4 to 5, 6 to 7, etc).

4.5   Graphs and Histograms

Graphs and histograms are used to see and communicate shapes, but not detailed numerical values.   The latter is better done by tables. (This limitation of graphs is picked up further in Chapter 17.)   Exercise 4.4 tries to clarify the basic notion of how histograms use unequal intervals.

4.6   Standardised Variables

Standardised variables are widely used in certain areas like psychometrics, but are not used much elsewhere.   But they also crop up in sampling theory (e.g. the t-ratio).

Pedagogically standardisation gives students practice in the concepts of means and deviations.   It may be worth introducing briefly at this stage, but can be left out if time presses.

4.6   Discussion

The formal sequence given in this section for summarising a set of data in terms of n, m, s, and the shape of the distribution may be worth spending a bit of time on.

It typifies the mathematician's way of "decomposing" a problem into separate independent factors: m for example does not depend on n; s does not depend on n or m; and the formula of the distribution does not depend on the numerical values of any of the parameters in it.

The number of readings n is generally not important (unless it is very small); hence we work with averages. When we then compare different sets of data (of a similar kind), the shape of the distribution and the amount of scatter is often much the same. The comparisons can thus be boiled down simply to differences of the mean values. (However, the size and nature of the scatter and the distribution shapes should at least be implied, even if not always stated explicitly.)

CHAPTER 5 : THEORETICAL DISTRIBUTIONS

The main role of theoretical distributions as described in this chapter is descriptive, i.e. to help us in comparing different sets of empirical data (e.g. "They are all Normal!"). Stochastic interpretations are outlined in Chapter 6. As noted in the introduction to the chapters, much of the detail does not have to be fully learned and remembered but is there for later reference.

5.1   The Normal Distribution

We need to stress the vast practical importance of the Normal because of its ubiquity and the fact that it is exceptionally simple since it always has the same basic shape (e.g. 68% of the readings always lie between $\pm 1s$). We really only need the percentiles for $\pm 1s$, $\pm 2s$, and $\pm 3s$ to characterise the distribution. A few percentiles like this are usually enough to let us check on any consistent biases across different observed distributions (e.g. to see if they are always a bit short in the lower $2\frac{1}{2}\%$ tail).

The formula for the Normal density function at the top of page 54 is given only because some statisticians feel it ought to be there. Personally I have not used it for at least 30 years. (The continuous nature of the Normal distribution is not greatly stressed since the Normal is often used as an approximation for discontinuous data.)

I have given the Normal distribution before the Poisson or Binomial (unlike some other texts) because it is simpler and much commoner in practice. But teachers can change the order.

5.2   The Poisson Distribution

I briefly note the basic assumptions for the Poisson to indicate that reasons can exist for such a distribution to occur. The crucial point is that if the observed data can be modelled by a Poisson, then the under-lying theoretical formulation may help us to interpret the nature of the data. This dual process of modelling and interpretation is relatively complex. Hence it is quite good to go over the interpretation twice, here and again in Chapter 6.

Figure 5.5 shows that Poisson distributions can have quite different shapes. We can say that this is typical of most theoretical distributions.

(The Normal distribution is almost unique in always having the same shape.)

Students cotton on to the fact that for a Poisson, $\mu = \sigma^2$. It is the sort of bald fact that is helpful. It also provides a quick check on observed data to see whether a Poisson fit is at all reasonable (Is $m = s^2$?)

Otherwise, the mathematical calculations of the Poisson need not be discussed much if the class are unlikely to need such expertise. In other cases one may want to emphasise them more. If so, one can mention the quicker recurrence formula for calculating p_r , namely that $p_r = (\mu/r)p_{r-1}$. Starting with the observed value of p_0 , one multiplies this by $(\mu/1)$ to give p_1; then multiplies that by $(\mu/2)$ to give p_2; and so on.

This will give slightly different values from the formula given on page 60 of the Primer. Students with a mathematical background find this upsetting, but it contains an important lesson: Different ways of calculating the theoretical distribution would give exactly the same answers if the fit were exact; but they will give (slightly) different answers if the distribution is only an approximation to the observed data. (E.g. the observed value p_0 with which we start the recurrence calculations is not exactly the same as the value given by $p_r = \mu^r e^{-\mu}/r!$ for $r = 0$.)

The Negative Binomial Distribution

The fact that for the Poisson $\mu = \sigma^2$ implies that the distribution can only fit a very limited range of data. Hence it is good to introduce more flexible forms of such distributions: the Negative Binomial Distribution is a common example. It fits data on the incidence of accidents, the distribution of plants or insects across space or time, the number of purchases of frequently-bought goods (as in Table 5.7), and so on. An example from the students' own subject-area would be super.

However, there is no need to go into the mathematical details of Table 5.8 in class. (The calculations are usually done on a computer; a numerical illustration is given in Section 12.3 of Data Reduction.)

5.3   The Binomial Distribution

Students sometimes feel uncertain about how the Binomial differs from the Poisson. It is therefore worth noting both the underlying similarities and the differences, as in the second paragraph of Section 5.3.

The remaining discussion of this section can be severely curtailed (i.e. cut) if time presses or if the class would find it too technical.

It may however be worth spending a little time on the fact that the mean and the standard deviation of a theoretical Binomial can be calculated in the usual way, as $\text{Sum}(x)/n$ and $\text{Sum}(x-\bar{x})^2/n$.

The theoretical frequencies in Table 5.11 can be used as a numerical illustration.   (Table 5.10 is not good to use because the high readings there are grouped.) Thus in Table 5.11, the mean of the theoretical distribution is $(5 \times 3.6 + 4 \times 17.1 + \ldots 0 \times 2.7)/100 = 2.575$ (the frequencies being percentages). This is $5 \times .515$ boys per family. The variance similarly is $\{(5 - 2.575)^2 \times 3.6 + (4 - 2.757)^2 \times 17.1 + \text{etc}\}$.

The theoretical formulae np and $\sqrt{\{np(1-p)\}}$ are only mathematical short-cuts.   They occur because the x's can only take two possible values, i.e. 1 and 0 (boy or girl), which simplifies the mathematics.   Thus for the mean, if the proportion of 1's (children who are boys) is p, then for the average family with n children, we can see that the mean number of boys must be n times p (or $5 \times .515$).

For a relatively mathematical class, the more formal proof that the average value of r is np comes from working through the sum of the observed values r times the proportion of times p_r that they occur. Thus from the Binomial formula for p_r on p.66,

$$r\,p_r = r \cdot \frac{n\,!}{(n-r)\,!\,r\,!}\ p^r (1-p)^{n-r}$$

$$= \frac{n\,!}{(n-r)\,!\,(r-1)\,!}\ p^r (1-p)^{n-r}$$

$$= np \cdot \frac{(n-1)\,!}{(n-r)\,!\,(r-1)\,!}\ p^{r-1} (1-p)^{n-r}\ .$$

Summing this over all r gives

$$np\ \Sigma\ \frac{(n-1)\,!}{(n-r)(r-1)\,!}\ p^{r-1}(1-p)^{n-r}$$

$$= np\ ,$$

since the summation is simply the expansion of the binomial $\{p+(1-p)\}^{n-1}$, which equals 1.

The variance formula $np(1-p)$ follows by working through $\Sigma\,(r-np)^2\,p_r$. This is hardly worth doing in class, even with a rather mathematical group, since it involves the same mathematical tricks rather than anything which would lead to a deeper understanding of the results.

The Limiting Cases

The first of the two "limiting cases" of the Binomial mentioned on page 68 refers to "rare events". Here the Binomial formulae become the same as the Poisson ones with a low mean. This is interesting, but of far less practical import than is sometimes implied. I would not stress it in class (or hardly even mention it) for an introductory course. Some books however give the mistaken impression that the Poisson applies only to cases with small means, i.e. to rare events.

The second limiting case, the Normal approximation to the Binomial, is also not specially important, since so many things are Normally distributed in the limit. (A technical problem over continous and discontinuous distributions arises here: The Normal approximation, strictly speaking, describes the proportion of readings lying in a certain definable interval around each integral value of r.)

The Beta-Binomial Distribution

This is one of the possible extensions of the Binomial Distribution. One can cut any detailed discussion, if necessary.

The main idea to stress is again that there is a range of distributions and some may give a better fit for data of a broadly Binomial kind (i.e. counts with an upper limit). The discussion of the BBD is taken further in Section 6.2 of the next Chapter in terms of the underlying stochastic model for the incidence of boys and girls.

The formula for the BBD in Table 5.12 is relatively complex and can easily be by-passed. But there it is, for the record.

If the class can take it at this early stage, it may be worth noting briefly that the BBD has <u>three</u> adjustable parameters: i.e. the fixed number of items n and the abstract quantities α and β, or - more concretely - n and two statistics like the mean and the variance.

With three parameters, the BBD is therefore more flexible than the other distributions discussed so far. There is for instance no fixed relationship between the variance and the mean - one can be greater or less than the other. The shape can also be more varied, taking not only the three forms of the Binomial distribution shown in Figure 5.6, but also a U-shaped form, as in the numerical illustration of Table 5.13.

A Background Note on the Binomial

The Binomial distribution can provide a model for a variety of observed data (e.g. on the incidence of faulty items in quality control, or the incidence of boys and girls in families of a given size), and not just for the much more limited sampling types of data discussed in many statistical texts (e.g. games of chance, or sampling from a quality control set-up).

To clarify the issue, I stress that the quality control example of Table 5.10 does not involve any sampling. It concerns the whole of a production run over a month, say, and the occurrence of faulty items in all the successive batches of 20 in which the items are parcelled up. No sampling has taken place. The data may (or may not) follow a Binomial distribution. Whether it does is empirical. It depends on the incidence of faults in the predetermined batches of 20 items. Does this follow an as if random stochastic process of the right kind (as discussed in Chapter 6)?

The situation would be quite different if one were taking random samples of n = 20 from the given production run, e.g. one sample of 20 each day. The distribution of the number of faults in these samples would then be Binomial whatever the underlying pattern of faults in the population, because of the random selection that has been deliberately imposed.

This issue should probably not be raised in class at this stage, where it would be likely to confuse students, especially since sampling has not yet been discussed. (One can raise the sampling applications of the Binomial in Part Three, although it is not explicitly treated in the main text there.) In my experience students have no problem with the idea that the Binomial is a possible model that can fit certain kinds of observed data.

5.4   Discussion

Theoretical frequency distributions are useful in the cases where they work. They then provide a succinct (if approximate) description of the observed data. But there is little merit in fitting a distribution to just one set of data. The pay-off comes when the same model describes many different data-sets, as in the blood sugar and breakfast cereals examples on pages 57 and 63. This is worth stressing, as students sometimes think that fitting a single data-set nas some mystical pay-off which they have failed to grasp.

For humped-backed symmetrical distributions, the usefulness of the Normal distribution lies in its simplicity (i.e. that it always takes the same shape – 68% between $\pm$ 20, etc). For skew data, we will have already stressed in previous class-sessions how difficult it is to provide succinct ad hoc summaries. Hence the theoretical distribution discussed here can be exceptionally helpful <u>when they fit</u>.

The Question of Fit

"Fit" is not a question of statistical significance here (except when one is dealing with sample data – and small samples at that – as will be discussed in Part Three).

In practice, one never expects an exact fit even for population data. So the real questions of fit are (i) whether the discrepancies are relatively small compared with the range of observed values, and (ii) whether there are consistent (i. e. "systematic") patterns in the discrepancies for different sets of a similar kind.

The distribution of boys and girls in families of a given size is a case in point here, as is taken up further in Section 6.2 of the next chapter. Here the fit is "close", i. e. discrepancies of less than 1 percentage point between the observed theoretical relative frequencies. But some discrepancies are "systematic", i. e. there are <u>always</u> slightly more all-boy or all-girl families than in the theoretical Binomial distribution.

Summary

The summary of the five theoretical distributions in Table 5.14 can be added to by reminding students that for the Poisson, Negative Binomial, and Binomial distributions the variances are respectively equal to the mean, less than the mean, or greater than the mean (as noted in the text). But for the Normal and Beta-Binomial distributions there is no general rule: the mean and variance can differ independently of each other.

CHAPTER 6 : PROBABILITY MODELS

As noted earlier (p.15), probability arguments are somewhat complex. Hence it seems best to introduce students to descriptive data-handling matters first. (In any case, most students make little use of probabilities.) But some teachers may like to introduce probabilities earlier. Chapter 6 can then still be used as background reading, although there are a few cross-references to Chapter 5.

Except when randomness is explicitly introduced as in random sampling, I believe that we can never know whether a physical process is really random or probabilistic. But if after extensive analysis the observed variation appears to be irregular, one can use probabilities as a possible "as if random" model. This can then be a useful descriptive device. It can also rule out various previously-possible interpretations of the observed phenomena. But one does not have to adopt any deep "causal" interpretation of randomness. (Teachers can put forward a different point of view here, probably without any major clash with what is said in Chapter 6.)

This approach implies that there are mathematical probabilities on the one hand (e.g. the addition and multiplication rules) and various empirical applications on the other. The latter differ from each other a good deal and do not have to obey the same philosophy, but merely use the same probability mathematics as a possible model.

6.1   Probability Distributions

As an introduction to probability distributions we can consider the case of children's heights. (A variable specifically from the students' subject-area would be more helpful.) Systematic differences in height have been established between boys and girls and by age, race, socio-economic background (possibly nutritional), inheritance (family similarities), and so on. But for any given sub-group of children, the variation in their individual heights can usually not be pinned-down more systematically. This kind of irregular "residual" variation is then described by frequency distributions. It can then be useful (but not essential) to describe it in probabilistic terms, treating the variation as if it were random. But this does not assume the individual variations in height are "really" random (whatever that might mean).

A second example one can discuss with the class is errors of measurement. Differences between repeated observations of the "same" phenomenon are not inherently random. But if there are no discernable systematic patterns, one may regard the irregular variation as being "as if random".

A third example arises with deviations from a model, such as a relationship between two variables, as discussed in Part Four. (I believe it is useful — if time permits — to look forward like this once or twice, introducing briefly one or two of the concepts arising later in the course. It helps to link things up this way, briefly at least.)

Thus one can draw on the board or overhead projector a scatter diagram for two variables x and y (using relevant labels for these from the students' subject-areas) and draw a "by eye" line through them. The (vertical) deviations of the dots from the line are then not inherently random. But if no systematic sub-patterns can be found, it can again be useful to summarise them as being "as if random". But this (again) does not mean that these deviations or errors are "really" random.

One can also point to the familiar example of tossing a coin twice. For an evenly-balanced coin the resulting theoretical probability of getting two heads is 25% of all such cases. Finding that it is about 25% in a long series of pairs of throws is then one check of the as-if-randomness of the observations.

6.2   Probability Processes

This section goes into probability processes as a statistical instance of the general processes of model-building and explanation in science. It uses "real" examples rather than games of chance. It may be curtailed or omitted if it is thought irrelevant or too advanced for the class.

A common misconception is that for a yes-no process to be "random", the chances have to be 50:50. The fact that boys tend to arise in 51.5% of births is a good counter-example to give. The incidence of boys and girls at birth need not be equal (the bias is actually towards boys), but the sex of each individual child can still be determined "as if random". It is like tossing a slightly biased coin.

The boy-girl example centres on the question of whether the likelihood of having boys differ from family to family. (If it does, there would then have to be some specific physiological or astrological cause). Initially such real differences look quite likely, since some parents do produce only all-boy and some produce only all-girl off-spring (and some kings have divorced their wives because they didn't/couldn't/wouldn't produce male heirs).

But the close fit of the theoretical "as if random" Binomial model with a common value of p = .515 rules the hypothesis out, given that it holds for so many sets of data (e.g. different size families, etc.). It shows that we could get the observed incidence of all-boys, etc, just by chance, without any "real" difference between different families.

However, the (small) systematic discrepancies from the Binomial then lead to the possibility that there is after all some (small) variation in the likelihood of different parents having boys. Another possible explanation would be the assumption of "independence" (i.e. that the sex of the first child does not affect that of the next one). This need not be so. Having a girl for a first child might for example influence the biochemistry of the mother to be subsequently more favourable to the survival of sperms with female genes. While the good fit of the theoretical Binomial model, which assumes independence, largely rules that out, the small systematic discrepancies in the fit might indicate some (small) lack of independence. That would have to be further explored (e.g. in terms of any special but small regularities in successive births, e.g. "runs" of 3 girls occurring more often than the Binomial would predict).

6.3   Games of Chance

Games of chance are well known. But they greatly differ from other situations in empirical science and everyday life. Thus the definition of probabilities in terms of "equally likely events" is not common elsewhere. What is more, it is rather "improbable" that coins with obviously different sides, or dice with different numbers of spots gouged out from each side, should have equally probable outcomes! As this kind of probability definition seems difficult and does not appear to apply to other scientific situations, it is not mentioned in the Primer, even though it has been around for centuries.

Games of chance do however allow us to illustrate the not "really" random nature of even such supposedly random phenomena and the need for establishing empirically whether they do behave "as if random".

6.4   The Central Limit Theorem

This serves to explain why something like a Normal distribution occurs so often. But observed data are usually only approximately Normal, even for very large samples or whole populations. Hence some of the assumptions in the theorem are not fully met in practice, which is reasonable enough. (Successive measurements may not be quite independent in the case of errors of measurement. Or in the case of people's heights, siblings' heights will tend to be correlated; but this is submerged when taking a large population of many different families.)

Students may need to be told that a proper proof of the theorem is difficult (it took more than 100 years for a "rigorous" proof to be developed). This is not surprising, since it involves going from some <u>qualitative</u> assumptions about independence, additivity, and large n to a very specific mathematical formula.   A very remarkable achievement!

But we can nonetheless explain (in a hand-waving sort of way) why something like the theorem should hold. We consider readings which are influenced by lots of different independent factors.   Some of these would make an observed value a bit bigger and some would make the reading a bit smaller (like the underlying causes of irregular errors of measurement). In most cases these various effects will roughly balance each other.   The resulting measurements will therefore be close to the true value, so that the observed distribution will have a large hump in the middle (at or near the true value).   But sometimes a number of positive factors will occur together giving a large positive deviation; and sometimes a number of negative factors happen to occur.   These more extreme readings will be rarer.   Hence something like the Normal distribution.

6.5   Subjective Probabilities

Subjective probabilities are worth mentioning briefly, because we use them colloquially in everyday life.   But they are seldom (if ever?) used for numerical calculations then.   (When have you, or somebody you know, actually done so? And if so, more than once?   I know nobody who has, and that includes me.)

The discussion can be elaborated in a subject like management studies, where the teacher should be able to produce additional reading material on decision theory (notes, duplicated papers, or chapters in other texts).

Prior Probabilities

We often have prior knowledge before analysing a given set of data. (This notion has been submerged in the last 50 years by Fisher's success in getting the most out of a single set of data.)

There are different ways of making use of prior knowledge: e.g. the design of a study; the choice of null hypotheses in tests of significance (Section 10.2); the fitting of a relationship (Section 13.3); the development and use of empirical generalisations (Chapter 19).

Yet a further way is in terms of the probability with which we hold some hypothesis before the new data are collected. This is formalised in the Bayesian approach. But Bayesian statistics do not appear to have produced much in the way or results or routinely-used procedures. How far to discuss it in an introductory course will depend on the individual teacher. (I personally have not found it useful; in my own analytic work, I make much use of prior knowledge, but it is in the ways noted in the preceding paragraph.)

6.6 Discussion

In summarising this chapter we can repeat that there are two ways to look at probabilities:

(a) As mathematical models; these must be right, but are just abstract statements.

(b) As describing observable phenomena, without reading anything deeply causal into it. (The occurrence is only "as if random".)

This pragmatic approach reduces problems that students have with probabilities. It also helps to remember that probabilities do not figure large in the rest of the course.

The Chapter 6 "Technical Note"

Those in the class who have sufficient mathematical background will like going through the technicalities of the multiplication and addition rules of probabilities on page 86 of the Primer. But there is not much point in teaching them in detail to other students (they are mostly unlikely to use the rules in their other studies).

The notion of <u>independence</u> is however well-worth stressing throughout this chapter. A common definition is that two events are independent if the outcome of one is not affected by the outcome of the other. But this is rather abstract. A more concrete statement is that two events are independent if <u>knowing</u> the outcome of one event does not help us in predicting the outcome of the other. (Typical examples are successive throws of a coin, the sex of successive babies born in a given family, or the sex of successive babies born in a given hospital.)

Part Three: Sampling

Any observed data involves some kind of selection. But usually it does not involve statistical sampling, let alone random sampling. I believe that we need to admit this to the class, possibly more than once.

Statistical sampling saves much time and money when we have a well defined but variable population. If we take a small sample, it introduces errors. These we must then allow for – hence statistical inference.

Dealing with errors is however negative, a matter of hygiene. It does not add anything positive to the data. It is not as good as if we had taken a really big sample in the first place. So while sampling and inference are of practical importance, I personally believe that they are not as fundamental as some statistics texts appear to make out. That is why I stress (§7.7) that "A sample can never tell us more than if we had measured the whole population." This may be a personal view but it can affect the tone of what we say about sampling and inference.

CHAPTER 7 : TAKING A SAMPLE

Here I stress to begin with how much of the sampling which we come across is not random.

7.1 Simple Random Sampling

It would be a good idea to illustrate the basic process of random sampling by an example from the students' subject-areas and to talk it through like the "homes with freezers" case in the text.

Random Selection

I believe that the word "random" should be used to represent the use of some formal random or probabilistic selection procedure, and not just for any kind of "irregular" or "haphazard" selection. Exercise 7.6 gives an illustration for students. Another (from a University exam) said that "a housewife goes to the freezer cabinet in a supermarket and picks out two packets of frozen peas at random". No she didn't: she did not

take all the packets out of the freezer cabinet, number them, put coded
slips in a hat, and shuffle these and select two. Instead, she selects two
packets more or less haphazardly (but not strictly "at random") from those
that look equally undamaged and are about equally accessible, i.e. near the
top of the cabinet. (Another splendid example of this kind is quoted in
Exercise 16C of Data Reduction.)

It is good to illustrate the potentialities of biased selection. For
example, I sometimes write up on the board the numbers

$$1, \ 2, \ 3, \ 4, \ 5, \ 6, \ 7 \ .$$

I then ask each student to select two representative numbers and to write
these down. General experience in collating these selections is that
the two end-numbers (1 and 7) are under-represented.

Students should also go through a practical exercise using random
numbers or pulling slips out of a hat (or the equivalent).

7.2   Cutting the Cost

The discussion here is in terms of sample surveys. This is of
some interest to most students (e.g. opinion polls). But in some subject-
areas there will be more suitable illustrations than selecting towns and
people.

A technical point noted in the text is that statisticians differ in their
definitions of multi-stage and cluster sampling.

Stratified sampling is only effective if the strata are really very
different in respect to the measured variables. Otherwise it does not do
much positive good. (But also does no harm.)

7.3   Probability Sampling

If this is not important to the particular class of students, the
section can be left out.

We can discuss here the perennial issue that people (rightly) feel
that the size of the population should affect the accuracy of a sample.
In sampling without replacement (which is what people would think of),
sampling errors for a given n do get bigger if the population is bigger.
But the numerical effect is tiny unless the population size N is really small.
Thus the accuracy of the sample mean (i.e. its standard error as in
Section 8.4) is affected by the factor $\sqrt{\{(N-n)/(N-1)\}}$. If N is a

million and n is a thousand, the effect is of the order of .999. So people are right in theory, but wrong (i.e. trivial) in practice.

Experimental Design

For classes where experimental design matters more than survey sampling, there is more discussion of randomised experimental design in Part Six (Section 21.3). But one may need to supplement the Primer with added specialist material.

7.4   The Population Sampled

The definition of the population to be sampled is crucial. Often people do not even know the size of the population which they are loosely considering. It is good to get the class to discuss some examples of how they would define populations in specific cases from their own subject-areas.

The big practical problem is building an up-to-date sampling frame (if none exists). Again, it is good to get students to discuss a case or two in class, or to construct a sampling frame as an exercise or as part of a project (Exercise 7.2 gives an example).

Non-sampling errors are also of vast importance − much more so than the space we give them in statistics texts (including the Primer).

7.5   Other Types of Sampling

Systematic sampling of lists is very common. This is all right if one is "pretty sure" that the list has no systematic pattern. (It is surprising how reluctant we are to use random numbers to make sure of avoiding bias!)

Quota sampling is sometimes too harshly condemned by statisticians, even though many short-comings in random sampling are lightly passed over (e.g. a response rate of only 40%). But well-validated forms of quota sampling will tend to give valid answers. (The usual statistical criticism is however not that quota sampling may be seriously biased, but only that one cannot calculate sampling errors! But even with probability sampling, how often are sampling errors calculated at all, or correctly estimated with complex designs, or really used?)

7.6   The Non-Sampling Approach

This is very important. It opens up the area of developing a generalisable finding − one which holds for quite different populations or samples therefrom – and the idea of empirical generalisation. It under-lies science generally and is further discussed in Chapter 13 and in Part Six. This subject is often not very developed in statistical texts.

Special attention can be paid to the non-sampling approach in the context of any student projects.

7.7 Discussion

Good sampling (including constructing sampling frames and reducing non-sampling errors) requires experience and skill. But all students need some appreciation of the general principles.

CHAPTER 8 : HOW SAMPLE MEANS VARY

Students often find sampling distributions difficult. This is not surprising. The very name "The sampling distribution of the mean" is not clear to newcomers; an alternative like "The distribution of sample means" is easier.

More important, the topic is theoretical and in an unfamiliar area. It therefore helps to stress that the theory is just a short-cut to what one could do empirically by simply collecting lots of real samples from the same population and seeing how much they differ.

A further problem is that the topic of how sample means vary is only an indirect way of answering the real question: namely how close to the true population value an actual sample mean is.

It helps to explain that if the means of different samples were to differ a lot from each other, any one sample mean would tend to differ from the unknown population mean. In practice of course we only have <u>one</u> sample. But we can tell much from its internal variability: it's a bit like having lots of small samples. Thus if the individual readings in the sample differ a lot, then the average of these readings will probably differ quite a lot from the population mean. If not, not.

The process is one of successfully pulling oneself up by one's boot-straps (i.e. just using the sample one has): It is quite clever and exciting – but not obvious to the newcomer.

To illustrate with an everyday example for students, if virtually everyone in a random sample has just one radio, then other samples will show the same thing and each sample will show pretty accurately what the population is like (everyone has one radio). But if people (or households) vary a lot in how many radios they have (some 0, many 1, a few 2 or 3, etc), different samples can give rather different results. Thus no single sample will probably reflect very accurately what the population is like unless the sample is pretty big.

8.1   A Series of Samples

If there is time, one may take the class through an example from their own subject-areas (rather than Corn Flakes in Southampton). A number of different samples can be randomly selected from a given set of readings, to see what the results look like.

8.2   The Shape of the Distribution of Sample Means

The basic result – that the shape is usually Normal – is extraordinarily simple. It makes the shape very easy to predict!

Practical illustrations are helpful (e.g. using the results of the sampling exercise in the previous section). But drawing balls out of urns should be avoided – it's not like real life.

8.3   The Mean of the Sampling Distribution

Some students find the basic result here self-evident. It is hardly worth disabusing them (e.g. that s is not an unbiased estimator of σ, even with (n-1) as divisor) since the problems of bias are usually small.

One may not want to go through the numerical illustration in this section in class. But one can mention that the illustration exists and that the result can also be proved algebraically (i.e. for n > 3).

An non-rigorous explanation for the mean would be to say that in a long string of random samples of size n, each reading in the population must in the long run come up equally often. Hence averaging all the samples must give the same result as the average in the population as a whole. This often seems to communicate quite well.

8.4   The Standard Error of the Mean

The standard error formula reflects that sample means tend to be closer to the population mean the less scatter there is in the individual readings and the larger the sample.

To explain the square-root effect qualitatively, one can say that taking more readings provides progressively less information. Thus in the Cornflakes example of Section 8.1, the sample of n = 10 households

3, 0, 0, 4, 2, 0, 0, 2, 0, 12,   Average: 2.3

already shows that the mean in the population is something like 2, 3, or 4, and not 20 or 40. Having a sample of n = 20, i.e. another 10, would only marginally improve the precision of that estimate. And so on.

If there are enough students with the necessary algebra, it may be worth outlining a proof. This is easiest for samples of $n = 2$. The mean of any two sample readings x_1 and x_2 is $(x_1 + x_2)/2$. The variance of the distribution of the means of different samples is then the average value of the squared difference of each such sample mean from the population mean μ, i.e. the average of

$$\left(\frac{x_1 + x_2}{2} - \mu \right)^2 .$$

This expression can be rewritten as

$$\left(\frac{x_1 - \mu}{2} + \frac{x_2 - \mu}{2} \right)^2 = \left(\frac{x_1 - \mu}{2} \right)^2 + \left(\frac{x_2 - \mu}{2} \right)^2 + 2 \left(\frac{x_1 - \mu}{2} \right) \left(\frac{x_2 - \mu}{2} \right) .$$

The average value of $(x_1 - \mu)^2$ over all possible sample values of x is by definition the population variance σ^2, so that the average value of $(x_1 - \mu)^2/4$ is $\sigma^2/4$. And the same for the average value of $(x_2 - \mu)^2/4$. As for the last term, since x_1 and x_2 were selected at random (i.e. independently), we can first calculate the average value for all those samples where x_1 takes some particular numerical value, so that $(x_1 - \mu)$ is constant. Then the average of $(x_1 - \mu)(x_2 - \mu)$ across all possible values of x_2 is zero (since the average value of x_2 is μ). And similarly for any other specific value of x_1.

Hence we get as the variance $\sigma^2/4 + \sigma^2/4 = \sigma^2/2$, or $\sigma/\sqrt{2}$ for the standard deviation. (The proof generalises for $n > 2$.)

Estimating $\sigma/\sqrt{n}$

Students usually think that estimating the population value of the standard error $\sigma/\sqrt{n}$ by $s/\sqrt{n}$ is a very obvious thing to do (once they are told that one can do so!)

The estimate must be less accurate than if one could use $\sigma/\sqrt{n}$ (since s has its own sampling error). But in practice the effect of using s is small (as will be seen in Section 8.5).

Interpreting the Standard Error

Students may raise two difficulties. One is that the standard error formula does not tell them what the actual sampling error for the particular sample is, but only how likely large or small sampling errors are. The answer is that that is already pretty clever. Their second difficulty is that they don't know how close the sample mean is likely to be to the population mean μ when μ is not known! That problem is dealt with in the next chapter (in terms of confidence limits).

8.5   Student's t-Distribution*

Using the estimated standard deviation s in the standard error formula $\sigma/\sqrt{n}$ should affect the quantitative conclusions one can draw. But the t-distribution shows that the effect is usually not at all big. As Table 8.2 implies, one need bother with it only for samples less than n = 10 or so (rather than n = 30 traditionally quoted in text-books). And very small but truly random samples are not common (why take only five readings when more are available?). However, t-values are widely quoted in the literature (in particular with multiple regression), and hence students need to know the general idea.

It may be useful to make explicit that in the t-ratio as such one is merely comparing the difference $(x - \mu)$ with the standard error of the difference, $(s/\sqrt{n})$.

Degrees of Freedom can now be introduced and related to the (n-1) divisor in the variance formula in Chapter 2. (Degrees of freedom will be further explained in §9.1 of Chapter 9 in terms of getting an unbiased estimator of s .) But for most students on an introductory course this seems like a great deal of fuss about nothing much.

8.6   Determining the Sample Size

This is a practical application of the standard error formula. As noted in the Primer, there is usually no simple, single answer. But because of the $\sqrt{n}$ effect - that markedly different sample sizes do not differ that much in their sampling accuracy - the exact sample size does not matter much. We can therefore be guided in practice by cost considerations, by the need for suitable sub-sample sizes, etc.

In a discussion at the Royal Statistical Society some years ago Professor George Barnard explained his personal approach as a statistical consultant to sample size questions as follows In his initial meeting with his client he would try to find out roughly the cost of a single measurement (e.g. the cost of an interview in a sample survey) and also the size of the available budget (but without making any of this too obvious). He would then ask for a couple of weeks time so as to appear to be earning his consultant's fee ("It's a difficult problem!") In the interval he would divide the budget by the cost per measurement, and then come back with

---

* It usually goes down well to tell students that Gossett was a master-brewer with the Guinness company in Dublin and was not allowed to publish under his own name. This was in 1908.

the answer ("A suitable sample size might be about 1200 or 1300" or whatever it came out to be). His client would always be very relieved ("That's just about feasible!"). And everybody lived happily ever after.

8.7   Other Sampling Distributions

The sampling distribution of the mean for simple random samples is very simple (unbiased estimator; Normal; standard error $= s/\sqrt{n}$). This simplicity is highlighted by problems that arise with other sampling distributions (like that of the standard deviation or a correlation coefficient) or for those of the mean when using sampling methods other than simple random.

At this stage students probably need nothing more than the "design factor" type of idea mentioned in the Primer.

8.8   Discussion

The basic ideas in this chapter are technically exciting, but probably less so for students with no major interest in statistics.

CHAPTER 9 : ESTIMATION

As statisticians, we know that problems of estimation can become quite technical. But the aspects that need to be covered in an introductory course are mostly simple and can be dealt with briefly.

9.1   Estimation

If we have taken a good (e.g. random) sample from a specified population and calculate the sample mean, standard deviation, etc., then these form the common-sense estimates of the population values.

It is not always quite so simple, as the (n-1) versus n kind of problem for the variance and the bias of s illustrate. But in an intro- ductory course one can just note that there can be different kinds of estimators with different kinds of properties, but that with reasonably large samples they tend to give much the same answer. (A complication not mentioned in the Primer is that estimators like maximum likelihood depend on specific assumptions about an underlying model, e.g. Normality.)

In the past, statisticians used to differentiate between point estimators and interval estimates. The latter in effect refer to what we now call confidence limits.

9.2   Confidence Limits

Since assessing the accuracy of a single sample is a bit of a miracle, it is not surprising that a logically precise statement about how close m is to μ (i.e. m's confidence limits) is really rather complex, the problem being that we do not know μ .

But we can again stress the common-sense side of it all. Thus faced by a sample mean of m = 24, say, with a standard error of $s/\sqrt{n} = 2$, it is roughly right to think that the population mean μ will almost certainly (i.e. 95% of the time) lie between 20 and 28, with the other 5% of cases mostly just outside (e.g. down to 19, or up to 29). This is all a bit loose, but not seriously misleading in practice.

I believe that it is often effective to give students something rather complex (as set out in the textbook) and then let them off the hook by stressing in class that the simpler common-sense approach will really do pretty well.

The Level of Confidence

Students need to be warned that the fuss they will sometimes come across about the precise probability levels (or P-values) is usually much overdone. Thus the numerical size of confidence limits for different probabilities, like 5% or 1%, differs very little (e.g. the 5% limits being 20 to 28 and the 1% limits 19 to 29 in the above example).

9.3   Small Samples

The main message for <u>very</u> small samples is that the confidence limits will be even wider than the small sample size would itself imply simply in terms of the standard error formula. It is here that the inaccuracy in estimating the standard error (i.e. using s instead of the unknown σ) does make itself felt – hence the t-distribution.

(Formulae for the confidence limits of small samples from highly skew distributions are generally not available because the mathematics are too hard.)

9.4   Prior Knowledge

The Bayesian approach has been widely discussed in recent years, but not much used in practice.

9.5   Empirical Variation

We will already have stressed the intellectual glory of assessing the accuracy of a single sample. But in practice we do not have to rely on this statistical feat all that much, because in science studies are usually repeated under different conditions to establish how far the findings generalise (see Part Six). This then also provides evidence of the maximum size of possible sampling errors.

9.6   Discussion

Basically this is a simple chapter. The hard work was in Chapter 8.

CHAPTER 10 : TESTS OF SIGNIFICANCE

Most students come across the notion of significant results in their other reading. But unless they become more involved in research and statistics, they seldom have to perform tests themselves.

The subject is a large one and is probably best taught selectively, emphasising aspects which are thought particularly relevant.

10.1 Testing a Statistical Hypothesis

Students need to become familiar with the basic argument: Does the observed result probably differ from the hypothesised value because of sampling error, so allowing us to ignore the discrepancy, or is this so unlikely that we had better reject the null hypothesis?

10.2 The Choice of Null Hypothesis

The hypothesis to be tested should usually stem from previous knowledge, i.e. reflect what we expect to happen.

This way of looking at sample data – testing the expected null hypotheses – is not widely discussed in statistical texts. But students do not usually find it difficult, and it seems in line with ordinary scientific practice.

The more traditional approach is to test a no-effect null hypothesis. But such testing has often been over-done in recent years and there is some reaction against it. (A no-effect test can be a safe-guard against over-interpreting our findings. But that shouldn't be our main aim in life. (In any case the ultimate proof comes from replication.)

There is a big difference between the two kinds of null hypothesis which we can emphasise: With a no-effect hypothesis a significant result is a "good thing" – hence the over-emphasis on "statistical significance". But with an expected null hypothesis, a significant result means that the analyst was wrong – the data are not as he or she had expected !

A technical point is that a null hypothesis has usually to be chosen before looking at the data. (It is OK to look at one's data and think of possible explanations; but usually we cannot then test their significance on the same data.   The technical procedures for doing so have mostly not yet been developed.)

10.3   Practical Significance

It helps to bring home to students that a statistically significant result only means that the observed result (or something numerically close to it) would probably have been found if the whole population had been measured.

It does not mean the result is in any way "important".   It may for example be quite small or inconsequential.

In no case does significance in itself mean that the result is generalisable.   It only says that the apparent result happened in the specific population that was sampled.

10.4   The Level of Significance

Many statisticians feel that too much fuss is made about the precise level of significance.   A 1% result should not be regarded as much "better" than a 5% result.   Either there is a real difference or there isn't.   A result cannot be "more significant" than that.

(There are cases in the twilight zone where there is a bit of doubt. But there is something peculiar about studies where that happens all the time.)

Over-emphasis on probability levels often has the effect that people do not look at how big the observed difference is or how it fits in with previous knowledge or expectations.

(As an example we can say that a large sample of people will un-doubtedly show the men to be "significantly" taller than the women, at the .0001 probability level or whatever.   But if the difference were only an eighth of an inch, that would be exceptionally small.   What matters is the size of a significant difference, not its probability.)

Type I and II Errors

The notion of Type I and Type II errors (or errors of the first and second kind) is good to explore with students.   It need not be restricted to their role in tests of significance, where they only matter in border-line cases.

Instead, we can briefly open up the whole business of the cost of alternative decisions (like the different costs of testing an aeroplane or other equipment for safety, or those of wrongly judging people guilty or innocent).

P-Values

P-values are a modern development not been mentioned in the Primer but which some students may come across.

Instead of reporting an observed difference as significant at a fixed probability level (say 5%), one would report the probability with which the difference (or a larger one) could have occurred under the null hypothesis, e.g. 7.2% or whatever ("P for probability"). This is meant to guard against over-emphasising a single rigidly applied cut-off point (like 5%).

But P-values seem to over-emphasise yet again the whole business of significance. Most results are either clearly significant or they are not. (Occasionally they lie in the twilight zone, in which case the study was wrongly designed !) The ultimate test is generalisation and replication in other studies, not just whether the result really happened in this one.

One-Tailed Tests

The idea behind one-tailed tests seems to make only limited sense. One may well expect a difference from the null hypothesis to be in only one direction. However, real observations in the opposite direction could still occur for some other reasons (e.g. that something went wrong with the experiment). In my experience one cannot rule out the possibility of alternative hypotheses in the unexpected direction.

A compromise approach might be to make the two "rejection tails" unequal, e.g. 4% in the expected direction (implying a lower criterion for calling a result in that direction "significant") and only 1% in the unexpected direction, adding to 5%. But that again over-emphasises the importance of trying to get "significant" results.

An alternative is to carry out an ordinary significance test with two equal tails. If the outcome should be about 5% (i.e. a bit dubious) but is in the direction of the "main" alternative hypothesis, one can be a little less dubious about it ! That will do , I reckon.

10.5   Tests for Means

Sections 10.5 and 10.6 outline specific tests for means and proportions, so that students can feel more confident when they come across applications in their other reading.

Working through numerical examples helps as a learning process. But I would hardly expect students to become very fluent with these procedures unless (and until) they were to use them much (e.g. if they specialise later in the more quantitative or empirical sides of economics or psychology, or whatever). A student project which actually involves random sampling and, hence, tests of significance would be the exception.

The really new material in this section concerns differences between two means (the class have already dealt with a single mean earlier). A point to bring out is that the appropriate standard error formula involves <u>variances.</u> This illustrates why these quantities got a special name earlier: they are useful in theoretical work. One can note that typically no corresponding theoretical formula can be worked out in terms of mean deviations.

A bigger point to stress is that even though the test is for the difference m_x minus m_y, the formula involves <u>adding</u> the variances, not subtracting them. One can point out that both $\overline{m_x}$ and m_y are subject to sampling error; hence the difference between them will have a <u>larger</u> sampling error than either of them on their own. That is why a plus sign rather than a minus sign "makes sense".

The derivation of the standard error formula $\sqrt{(s_x^2/n_x + s_y^2/n_y)}$ is fairly easy to sketch in. It may be worth giving in class if there is time and a sufficient proportion of students with some algebra (telling the others to "switch off" for a moment).

The standard error of $(m_x - m_y)$ is the standard deviation of the distribution of the difference $\{(m_x - m_y) - (\mu_x - \mu_y)\}$ across all possible samples of x and y of size n_x and n_y. The standard deviation is the square-root of the variance of the expression.

To get at this variance we can rewrite $\{(m_x - m_y) - (\mu_x - \mu_y)\}^2$ as $\{m_x - \mu_x) - (m_y - \mu_y)\}^2$. Multiplied out, this gives

$$(m_x - \mu_x)^2 - 2(m_x - \mu_x)(m_y - \mu_y) + (m_y - \mu_y)^2 .$$

The first term $(m_x - \mu_x)^2$, averaged across all possible samples of size n_x, is by definition the squared standard error of m_x itself. (This may need to be spelled out slowly to let students catch up on the idea again:

How much does m_x on average differ from μ_x, ...) It can therefore be estimated as usual from the n_x individual readings x in our one and only sample, i. e. as s_x^2/n_x. This is the first term in the standard error formula that we are trying to prove.

The third term in the above expression is similarly s_y^2/n_y.

The middle term $(m_x - \mu_x)(m_y - \mu_y)$ averages out at zero, because the x and y samples are selected independently. This means that there is no general tendency for a high m_x to go with a high m_y. (More precisely, for a given m_x, the average of $(m_y - \mu_y)$ across all possible y samples is zero – the same trick as on page 36 of the Guide.)

If the null hypothesis is that the x and y populations have the same distributions (not just the same means, i. e. $\mu_x = \mu_y$), it would make sense to calculate the standard error of the difference of the two means from a pooled estimate of the standard deviation of the individual readings, with a gain of one degree of freedom in the sensitivity of the test. That is why pooled estimates of the standard error are mentioned in the text (p.141). It seems a bit fiddly for an introductory course, but I have seen it in examination papers for first year students in subjects like psychology and marketing !

Paired Readings

The "design of experiments" concept involved with paired readings is very important in its own right, quite apart from sheer significance testing. If two measurements are taken on each "subject", or measurements on different subjects are paired by some stratifying factor, this can lead to much smaller sampling errors for the average difference between the x and y readings.

If there is time, one can work through the numerical example of Table 10.4 in class, treating the two sets of five readings as not paired, to show how much bigger the standard error of the difference of the two means would be. (Or one can set this as an exercise and briefly discuss the results in class.) For the x's on their own, the standard error of the mean of 20 is influenced by the large variation from 6 to 38; similarly for the y's (from 0 to 19). The variation in the paired differences (x - y) is less, from -1 to 9.

This topic is worth lingering on. (It should perhaps be expanded in any future edition of the Primer and linked to Part Six.) The design and interpretative aspects could easily figure in any empirical project work students do.

Analysis of Variance

Some statisticians (like myself) with fairly modern statistical training may have to adjust themselves to the fact that randomised experiments and the associated analysis of variance procedures, brilliant though they are, have relatively few practical applications.   The general ideas are nonetheless worth mentioning to students and may have to be elaborated for students who are more likely to meet up with them extensively in their other work.

It can help to say that the analysis of variance is so named because its technicalities involve decomposing the overall scatter (i. e. <u>variances</u>). But as noted in connection with Chapter 3, from a straight interpretative point of view the analysis should really be called the analysis of <u>means</u> (i. e. main effects and interactions).

Non-Parametric Tests

These have recently become rather fashionable amongst some statisticians.   Some old-established cases arise in §10.6.   But the scope for other important applications seems to be limited.

Thus students in most subject-areas only need be introduced to this topic in general terms.   They need to know the advantages (i. e. papering over some of the gaps in tests of significance for non-Normal distributions) and the limitation (not being able to describe what the difference in question actually is).   The treatment in class can easily be expanded if desired.

10.6   Tests of Proportions

Students may need to be told that the standard error formula for a percentage only applies to a variable which is classified into two categories (e.g. yes and no, male or female, buyer or non-buyer, etc) and not to other kinds of percentages (e.g. the proportion of sales of product X going to a certain region).

The mystery can be taken out of the $\sqrt{\{p(1-p)/n\}}$ formula by reminding students that (as already noted in Chapter 5) $p(1-p)$ is a short-cut to the usual variance formula $\mathrm{Sum}(x-\bar{x})^2/n$.   Thus if x is scored 1 in np cases and 0 in $n(1-p)$ cases, then $\bar{x} = p$, so that

$$\mathrm{Sum}(x-\bar{x})^2/n = \{(1-p)^2 np + (0-p)^2\, n(1-p)\}/n$$

$$= np(1-p)\{1-p+p\}/n = p(1-p).$$

(This is simpler to express in terms of p and q, but for many classes it is better to use (1-p) rather than introduce a new symbol q. However, q still needs to be mentioned, since students may come across the standard error formula as $\sqrt{pq/n}$.)

A special feature of the standard error formula is that it depends only on the proportion of yes's or whatever, i.e. on the mean p. Thus if we know or expect p to be a certain value, we know its standard error.

Contingency Tables

We can compare two proportions (e.g. the percentage of defective items produced by two different machines) by using the two corresponding values of $\sqrt{(pq/n)}$ in the formula for the standard error of the difference of two means (Section 10.5), if the n is large enough for normal approximations to hold. But usually such comparisons are treated as 2 × 2 contingency tables, as illustrated in Table 10.6. This approach can of course be extended to k rows and r columns.

Tests of significance for contingency tables are generally restricted to testing the null hypothesis that the classifications are independent (in the case of Table 10.6, that the proportions of defectives in the first machine and the second machine are the same). As with other non-parametric tests, if the result is significant one still has the task of describing the relationship. This can get complicated with larger tables.

Tests of contingency tables are only needed if the data are random samples from some large populations or if some explicit randomisation has taken place (e.g. in allocating patients to different treatments in a clinical trial). Otherwise one is really dealing with a small population and any observed difference is "significant" (i.e. real) but may of course be negligibly small or unimportant.

If students need to be familiar with larger contingency tables, it may be worth going through an example or getting students to do one. In any case, it needs to be said that the "expected" value for any particular cell is calculated as

$$\text{Expected value} = \text{Column total} \times \frac{\text{Row total}}{\text{Grand total}}.$$

I.e. the expected value is the proportion of that row total to the grand total, as applied to the column total (or vice versa, since the calculation is symmetrical between rows and columns). This was omitted from the Primer.

A brief discussion of the theoretical background may also be worth-while. First, these tests are a long-standing case of non-parametric tests: no assumption of Normality and no use of the standard error concept. Second, the actual sampling distribution of the calculated Sum $\{(O-E)/E\}$ is a hyper-geometric distribution which is nasty to work out numerically. But it can be approximated by the numerically much simpler chi-squared distribution, as set out in Table 10.8. (This approximation only works well if the cell expectations are greater than 5, hence that restriction.)

Goodness of Fit

Students will not come across goodness of fit tests much (if at all) in their other reading. Nonetheless, such tests can be briefly touched on, for three possible reasons:

(a) They may be required by the syllabus.

(b) In Part Two we spent time on fitting possible probabilistic or descriptive models to observed frequency distributions. With sample data (especially with small samples) we may want to test whether any discrepancies are real.

(c) They are another case of non-parametric tests (in as far as we ignore the specific features of the fitted model).

10.7   Discussion

Students will probably need to be warned again about "hunting", i.e. inspecting sample data and picking out the biggest difference, and then "testing" it, as mentioned in the Introduction. Traditional tests of significance will then vastly overstate the real significance of such effects, especially with small samples.

Part Four: Relationships

Everyone is interested in relationships. In data analysis they allow us to predict and to control; they can even lead to causal understanding. But what we teach about them in statistics – correlation, regression, and multi-variate methods – does not seem to be very effective. There are few, if any, obvious, well-established success stories.

Our traditional statistical methods are more limited than is widely thought. For instance, correlation and regression analysis apply essentially to only a single set of data. They do not deal with different sets of data .Nor do they claim to lead to generalisable results.

In Chapters 11, 12, and 14 the Primer tries to face up to the problems involved by first giving a brief warning at the beginning of each chapter that there may be difficulties with the traditional methods, then describing the methods, and finally concluding with a brief critique.

Chapter 13 discusses the alternative problem of fitting an equation to two or more different populations (or samples therefrom). This is not usually covered in introductory texts. It is not a new method, but makes explicit what scientists do in working towards the ordinary lawlike relationships of science. Students generally find no conceptual difficulties with this chapter. But if time or other considerations demand it, the chapter can be excluded from the main classroom syllabus and left for students to read on their own.

CHAPTER 11 : CORRELATION

The correlation coefficient is widely used. But it is difficult to know what any non-zero result really tells us. What does a correlation of .6 really mean?

The coefficient does not say what the relationship actually is, i.e. how y varies with x. As discussed more in §11.4, it only says that a relationship exists and roughly how strong it is. The trouble is that r as a single number cannot by itself tell us all we want to know about the relationship. Yet in a lot of studies only the value of r is reported.

In the classroom we can say that r tells us whether x and y are associated, but start commenting on its interpretative problems earlier rather than later. This will stop students worrying that they do not understand something when it is probably something we do not understand either.

11.1 The Formula

Having shown students some pictures of correlated data, we can say that various indices have been invented over the years to try to summarise such patterns, and note that r is the commonest in use. (The divisor (n-1) in the covariance formula is once again a rather irritating minor technicality which hardly matters here, numerically or otherwise – it is really just an average.)

To show that r makes sense as a measure of association, it is good to illustrate the four cases on page 158 numerically. We can take the means to be $\bar{x} = 5$, $\bar{y} = 25$ (as in the numerical example of Table 11.1) and consider some specific values of x or y either greater or smaller than these means, namely

(i) $x = 10$, $y = 30$. Here $(x-\bar{x})(y-\bar{y}) = 5 \times 5 = 25.$

(ii) $x = 3$, $y = 20$. Here $(x-\bar{x})(y-\bar{y}) = -2 \times -5 = 10.$

(iii) $x = 10$, $y = 20$. Here $(x-\bar{x})(y-\bar{y}) = 5 \times (-5) = -25.$

(iv) $x = 3$, $y = 30$. Here $(x-\bar{x})(y-\bar{y}) = -2 \times 5 = -10.$

The covariance, the average of all cross-products $(x-\bar{x})(y-\bar{y})$, will then be close to zero if positive and negative values balance out in terms of both number and size. (In the numerical example the cross-products (i) and (iii) are numerically much larger than the other two.)

One can now also use this numerical illustration to show how the value of the covariance depends on the units of measurement of x or y. Thus if all the x's were multiplied by 3 (as in going from yards to feet), the four cross-products in (i) to (iv) would each be 3 times as large, i.e. 75, 30, -75, -30 . But the standard deviation of the new x's would also be three times as big. Hence dividing the covariance by the standard deviation of x (and similarly by the standard deviation of y) gives a coefficient which is independent of the units of measurement of x or y. It is still zero when the covariance is zero (i.e. "no correlation"), and has a maximum of =1 and a minimum of -1 .

The last property of course refers to situations where all the observed points lie exactly on a straight line, if plotted on a graph, with a +ve or -ve slope. (If students want a proof of the ±1, one can develop it from saying that $(y-\bar{y})$ is then directly proportional to $(x-\bar{x})$, e.g. $(y-\bar{y})$ = $+k(x-\bar{x})$, and work through the covariance and the sd's from there.)

The short-cut formula for working out the covariance on page 161 is of the same kind as the short-cut formula in Chapter 2 for working out the variance.

11.2   Sample Data

With sample data, students may ask about testing against non-zero hypotheses. In a bivariate Normal population, the quantity $\frac{1}{2}\log_e\{(1-r)/(1-r)\}$ approximately follows a Normal distribution with mean of $\frac{1}{2}\log_e\{(1 +\rho/(1 -\rho)\}$ and variance $1/(n-3)$. (This was shown by Fisher in 1915, in one of his earliest papers.) But such a test is not often used because of the difficulties of interpreting a non-zero correlation coefficient discussed in Section 11.4.

11.3   Rank Correlation

Ranked data illustrate how it is not too difficult to invent different kinds of correlation indices. (The problem is knowing what they mean, other than for zero or 1.)

For students who might have to use rank correlation coefficients, one can note that the ordinary product-moment or relation coefficient, here called Spearman's ρ, can also be calculated by using a short-cut which involves only the differences d between the two ranks of each item: thus $\rho = 1 - \{6(\text{Sum } d^2)/n(n^2 -1)\}$. For the numerical example in Section 11.3, this is $1 - \{6(1+1+1+1+0)/5(25-1)\} = .8$, just as by direct calculation, but quicker. (Students should probably be told that this ρ is of course not the same as the ordinary Pearson ρ for the population.)

11.4   Interpreting a Correlation Coefficient

Reducing a scatter diagram to a few summary figures is a good idea (like the slope and intercept constant of an equation, plus a measure of the scatter about it).   But reducing the data to just a single summary statistic – a correlation coefficient – is overdoing it.   The difficulties of interpreting r stem from this.

The coefficient can tell us about the scatter (or strictly, the relative scatter) about a relationship – the usual $s_y^2(1-r^2)$ formula which is discussed in Section 12.3.   But r cannot tell us what the relationship actually is, e.g. whether the slope is steep or flat.

It needs to be rammed home that one can get a high r with a very low slope – a high r only means that the scatter of the readings is relatively small.   When people say that two variables are "highly correlated", they usually mean the correlation coefficient is high and the scatter relatively small, not that y varies greatly for each unit increase in x.   (This is discussed in Chapters 12 and 13.)

It is important to bring out these difficulties of interpreting r. Otherwise students are left with a feeling that they have failed to understand something.   (Students usually do not mind being shown the limitations of a statistical procedure!)

11.5   Correlation is **not** Causation

At this stage it is probably enough to stress this point with some further illustrations.   One could be the numerical example of supermarkets and small stores in Table 11.1 (one does not cause the other).

Another could be children's heights and weights as discussed in Chapter 13.   These vary together (i.e. correlate) as children grow, but height does not cause weight, nor vice versa.   Nor is it meaningful to consider that changes in height cause changes in weight, or vice versa. They are <u>associated,</u> but this says nothing about the causation.

Other examples from the class's main subject-areas would be very good here.   (Causation will be touched on in the next two chapters and is discussed further in Chapter 20 – e.g. that in Boyle's Law $PV = C$, pressure **P** does not cause volume V, etc.)

11.6   Discussion

Correlation coefficients can be of some use (e.g. the $s_y\sqrt{(1-r^2)}$ formula for the rsd in Chapter 12).   But on their own they are of far more limited use than appears to be generally recognised.   Over-indulgence in r's has probably held back quantitative social science a good deal.

CHAPTER 12 : REGRESSION

Least squares regression has been a standard technique for more than a hundred years. It provides a certain best-fit solution for a single set of data. But it does not lead to the lawlike relationships of science (as discussed in Chapters 13 and 19). Nor indeed does it aim to do so.

We need to teach regression because it is widely taught and used. Students often have some difficulty with it, probably because of the lack of a unique solution (the two regressions, y on x and x on y). I feel very dubious about it myself: in more than 30 years practical analysis I have not seen a case where the results have stood the test of time, including in my own work .

12.1 A Straight-Line Equation

In explaining straight lines we need to stress that the slope is constant, i.e. the same anywhere along the line (unlike a curve). Hence we can obtain the numerical value of the slope by taking any two points on the line and dividing the difference of the two y-values by the difference of the two x-values. Taking the points far apart on a graph reduces measurement errors.

Finding the mathematical equation which represents a line that has been drawn on a graph is a very different problem from deciding what line to draw through scattered points on a graph. Students sometimes confuse the two things.

The Fit of the Equation

The question of vertical versus horizontal deviation which arises here (e.g. "plotting the variables the other way round") is pretty deep. We return to it in Section 12.7.

12.2 A Best-Fitting Line

It is probably worth noting that on common-sense grounds one would aim to put a line through the means of the data. Otherwise there would be consistent bias in the deviations, e.g. more points lying above the line than below. So the real question is determining the slope.

To ram home the trial-and-error approach, students can calculate the rsd for the second line mentioned in the text, i.e. $y = 12.5 + 2.5x$:

						Average
Observed x	2	3	5	7	8	5
Observed y	20	18	27	26	34	25
$12.5 + 2.5x$	17.5	20.0	25.0	30.0	32.5	25
$y - (12.5 + 2.5x)$	2.5	-2.0	2.0	-4.0	1.5	0

The Theoretical Regression Formula

There is no great need to prove the theoretical formula $b = \text{cov}/x, y)/\text{var}(x)$, for three reasons. First, it is merely a short-cut. Second, it has been known for over a hundred years. (So why prove it again?) Third, non-specialist students are not greatly interested in statistical proofs (but usually find an explanation like the trial-and-error approach helpful).

If we want a proof, the traditional approach is through calculus. But a more helpful alternative comes from noting that the mean squared deviation from any straight line $y = a + bx$ is

$$s_y^2 (1 - r^2) + (bs_x - rs_y)^2 + (\overline{y} - a - b\overline{x})^2 .$$

(This can be derived from first principles by expanding the mean square $(y - a - bx)^2$. See also "How Good is Best?" J. Royal Statistical Society, A, 1982.)

The first term $s_y^2 (1 - r^2)$ is the same whatever the coefficients a and b. It is the familiar residual variance for the regression of y on x.

The second term depends on the slope b. It is least (i.e. zero) if b is chosen to equal rs_y/s_x or $\text{cov}(x, y)/\text{var}(x)$, i.e. the slope-coefficient of the regression of y on x.

The third term also depends on the intercept-coefficient a. It is least (i.e. zero) if a is chosen to equal $(\overline{y} - b\overline{x})$. It is the familiar intercept-coefficient of the regression of y on x, if b is rs_y/s_x.

This explanation can help students to see why the regression slope gives the least scatter.

12.3   The Residual Scatter

One often needs to compute the individual residuals from an
equation (or plot the data graphically) to check for any systematic sub-
patterns.   Table 12.1 and the example for $y = 12.5 + 2.5x$ have illustrated
how one can then calculate the rsd.

But there is also the traditional formula $\text{rsd} = s_y\sqrt{(1 - r^2)}$.   This
is another short-cut, to avoid computing the residuals.   It works simply
from quantities already calculated from the x's and y's, such as the
variances and the correlation r or the slope-coefficient b.

The formula is often interpreted in terms of variances.   Thus r^2
is regarded as the proportion of the observed variance s_y^2 which has been
"explained" or "accounted for" by x.   But this interpretation is numerically
misleading.   Variances do not describe scatter directly, unlike a standard
deviation which says that for a Normal distribution, 68% of the readings lie
within $\pm s$ of m.

One can illustrate the point with a specific value of r, say $r = .8$.
This is usually regarded as rather high and the r^2 formula says that as
much as 64% of the variance of y has been explained.

But the residual standard deviation of the y's is $s_y\sqrt{(1 - r^2)} = .6s_y$.
This is 60% of the originally observed s_y.   Thus even with an r as high
as .8, the residual scatter is still 60% (i.e. more than half) of the original
scatter.   Unless a correlation is very high indeed, the reduction in scatter
by a regression equation is small (as shown in Table 12.2 in the main text).

12.4   Sample Data

This section gives the formula for the standard error of the
regression slope.   It can be used for setting confidence limits, or for tests
of both zero and non-zero null hypotheses.

Some textbooks also give a standard error formula for an individual
prediction y from x.   But this is a relatively complex topic and is probably
best by-passed in an introductory course.

12.5   Non-Linear Relationships

A curved relationship is worth fitting if different sets of data show
the same kind of pattern.   The aim is to reduce the description of the data
to a systematic equation plus irregular scatter (which is then easy to
summarise).

Students on an introductory course would not be expected to develop such relationships successfully.   But they need to be aware that most relationships in science are non-linear.   They should learn to be comfortable with non-linear equations and the idea of transformations.   This is further discussed in Section 13.4.

12.6   How Good is the "Best-Fit"?

The "best-fitting" line does not give a much better fit than any other line that might reasonably be chosen (like most of those one might "draw by eye").   This came out in the trial-and-error approach in §12.2.

The formula for the mean squared residual deviation about any line $y = a + bx$ in Section 12.2 also tells this story.   Thus for any line through the means $\bar{x}$, $\bar{y}$, of the given data, the variance of the residual deviations is $s_y^2(1 - r^2) + (bs_y - rs_x)^2$.   This becomes $(1 - r^2) + (b - r)^2$ if for simplicity we take $s_y = s_x = 1$.   Then for a correlation of $r = .6$, the regression-slope b would also be 0.6 and the rsd $\sqrt{(1 - .6^2)} = .8$.

But a line with a 50% higher slope of 0.9 would have an rsd of only $\sqrt{\{(1 - .6^2) + (.9 - .6)^2\}} = \sqrt{.73} \doteq .85$.   This is only about 6% more than .8 .

Similarly, a line with the 50% lower slope of 0.3 would have an rsd of .85.   Hence, lines with slopes differing by up to 300%, from 0.3 to 0.9, would have rsd's at most 6% bigger than the least squares regression.

One needs to warn students about this.   It can stop them (and us) from reading too much into a regression solution.   (One can point out that the approach in Chapter 13 can give much more decisive results.)

12.7   The Regression of y on x and x on y

It has long been known that least squares does not give a unique answer: Minimising the residuals in the y-direction and in the x-direction gives different answers.   The slopes are $cov(x, y)/var(x)$ and $cov(x, y)/var(x)$ and one is not the inverse of the other.

If students ask why this is so, one answer is why not?   There would have to be a very special reason why the two ways of fitting a line should give the same answer !

Students often wonder whether minimising the squared perpendicular deviations would solve the problem.   This would give another line, the slope being

$$\frac{(\frac{s_y}{s_x} - \frac{s_x}{s_y}) + \sqrt{\left\{(\frac{s_y}{s_x} - \frac{s_x}{s_y})^2 + 4r^2\right\}}}{2r} .$$

But changing the units of measurement of y or x by a factor k does not change the slope by k. Hence this type of line is no use, as noted in the Primer. (A numerical example is given at the end of this chapter.)

We need to tell the students that there is no controversy about there being two regressions, and that there is generally no clear-cut or agreed resolution of the problem of how to choose between them.

We also need to introduce the idea of independent and dependent variables. But in many situations the distinction is not clear-cut. (I find it usually begs the whole question because relationships are not directly causal anyway.)

Another traditional idea is to choose the regression according to the way one wants to do predictions, y from x or x from y. But there are many situations where one does not want to "predict" at all. (When did you last use a regression equation to predict y from x in a real-life situation, rather than just talk about it?)

Instead, we often need to see whether the same observed x/y relationship holds again in the new data (which is a symmetrical statement about the two variables). The whole notion of prediction is usually hardly discussed in statistical texts (e.g. whether one is predicting for other readings from the same population or to other populations; see for example p.251 of the Primer).

It seems to follow that using or interpreting regression equations usually involves rather personal or subjective decisions. Although the regression formulae may appear very objective, their application often is not.

A special case where these various problems do not arise is when one variable, x say, is controlled at fixed values. This occurs in certain experimental situations, or when different groups of items or people are deliberately selected according to their known value of x (e.g. children of different ages). This type of case is further discussed in Section 13.5 of Chapter 13.

12.8   Discussion

Regression analysis is surprisingly unsatisfactory – lots of promise (the "best-fitting line"), but also apparently unresolvable problems. W.G.Cochran said long ago it was the worst-taught part of statistics.

One of the worst aspects is that the best-fitting regression equation for one set of data will generally not be the best-fitting equation for any other set of data (item (c) on page 182 of the Primer). Hence it is not surprising that no lasting empirical results have been reported.

Students may object to being bothered with all of this on an introductory course. This can be countered on three grounds :   .

  (i)   Regression is widely used, so students need to
        know about it.

  (ii) Many scientific subjects have unresolved difficulties
        and controversies. (Students were warned of the
        problems with regression at the beginning of the
        chapter.)

  (iii) An alternative approach exists – a different
        solution to a different problem – as described in
        Chapter 13.

Appendix on Perpendicular Deviations: A Numerical Example

We can use a numerical illustration to illustrate how the line which minimises the squared perpendicular deviations depends on the units of measurement.

Consider data with $r = .8$, $s_y = 1$ and $s_x = 2$. Then the slope of the regression of y on x is

$$\frac{rs_y}{s_x} = \frac{.8 \times 1}{2} = .4 .$$

If the units of y are made 10 times as small (as in going from centimetres to millimetres), s_y increases 10-fold from 1 to 10, and so does the slope of y on x,

$$\frac{rs_y}{s_x} = \frac{.8 \times 10}{2} = 4 .$$

Thus in the two regression equations of y on x, y increases by 4 millimetres or .4 centimetres for each unit increase of x, which is consistent.

The line of the form y = a + bx which minimises the squared perpendicular deviation has a slope of

$$\frac{(\frac{1}{2} - \frac{2}{1}) + \sqrt{\{(\frac{1}{2} - \frac{2}{1})^2 + 4 \times .8^2\}}}{2 \times .8}$$

$$= (-1.5 + 2.2)/1.6 = .44 .$$

If y is expressed in millimetres so that $s_y = 10$, the slope is

$$\frac{(\frac{10}{2} - \frac{2}{10}) + \sqrt{\{(\frac{10}{2} - \frac{2}{10})^2 + 4 \times .8^2\}}}{2 \times .8}$$

$$= (4.80 + 5.06)/1.6 = 6.2 .$$

which is not 10 times the "centimetre" slope of .44 .

CHAPTER 13 : MANY SETS OF DATA

Chapter 13 in the Primer concerns lawlike relationships: results which hold for many different data sets under a variety of conditions. The topic is not usually covered in statistical teaching or texts, but students do not find it difficult. If the topic is regarded outside the required syllabus or time presses, the chapter can be left to students to read.

Between Group Analysis (BGA) treats a different problem than least squares regression. The latter finds a best-fitting line for a single set of data. BGA finds an equation which holds for many different sets of data. In this approach many of the problems of least squares regression disappear, like the choice between y on x and x on y. The analysis is symmetrical in x and y, that is how the means $\bar{x}$ and $\bar{y}$ vary together between the different data sets.

13.1 Between Group Analysis

In fitting a straight line, choosing a line for a given set of data is no longer a problem once we have a second set of data. There is then only one straight line which goes through the means of both sets of data. If the means of further sets of data lie on the same straight line, that remains the line to fit.

In practice, the means $\bar{x}$, $\bar{y}$ of different sets of data will not lie exactly on a straight line (or on any simple smooth curve). This is not just due to sampling variation, but tends to occur however large the samples. (The reason is that there are also other variables in the situation.) But if the means lie roughly on a straight line (or simple smooth curve), one fits such a straight line or curve as a simplifying description.

If we force a straight line onto non-linear data we have to accept some ambiguity about precisely which straight line to use. The ambiguity of the initial working-solution tends to be reduced as more data become available, especially data outside the initial range of variation . This is illustrated further in § 13.4.

Students may ask why one does not then use some kind of best fit procedure (e.g. least squares). But this would not provide a unique solution (e.g. the y on x or x on y problem), nor would the data satisfy the usual least squares conditions (e.g. Normal distributions). The result would also depend on how many readings there were in each set of data.

If a clear-cut relationship exists the relevant correlation is much
higher in BGA than when faced with a single set of unstructured data. It is
the correlation between the $\bar{x}$'s and $\bar{y}$'s, not that between the individual
readings x and y, which affects our judgment of fit here.

13.2   Fitting a Working Solution

No short-cut mathematical formula is needed in BGA, and hence the
procedure may seem less complex or "mathematical" than least squares
regression. But obtaining the slope of the equation from the highest means
$(\bar{x}_h, \bar{y}_h)$ and the lowest means $(\bar{x}_l, \bar{y}_l)$ involves an objective equation, i.e.

$$\text{Slope} = \frac{(\bar{y}_h - \bar{y}_l)}{(\bar{x}_h - \bar{y}_l)} \quad .$$

For BGA to work, the different populations (or samples therefrom)
must have different means — the more different the $\bar{x}$'s (and the $\bar{y}$'s) are
from each other, the more powerful will be the resulting generalisation.
If the scientist or observer cannot create or select data sets with different
means, the relationship between x and y will be difficult to investigate.
The failure is empirical.

13.3   Prior Knowledge

A big difference of BGA from classical regression analysis is that
BGA usually makes direct use of prior knowledge when faced with a new set
of data, instead of always starting from scratch.

In most areas of study, similar data have been analysed before.
We have to ask "What is already known" and see whether the previous
results fit again. If it does, it is a further extension of the existing
generalisation.

But we need to emphasis to students that this may fail: the new data
may not be like the previous data ! A real-life example is discussed in
Section 6.7 of my 1975 Data Reduction. (In the initial printing of Data
Reduction Figures 6.1 and 6.2 were transposed.)

13.4   Curved Relationships

The nature of curved relationships is discussed more fully here
than in the regression chapter, because students are by now more familiar
with the nature of relationships and should therefore cope better.

The height and weight illustration shows how ambiguity over which equation to fit decreases as more data become available. (This example is discussed more fully in Chapters 7 and 8 of Data Reduction.)

Table 13.3 shows that powerful results can be obtained from just a few studies. It also shows that lawlike relationships typically do not hold universally. Much of the work involved usually lies in systematising the exceptions (e.g. older girls).

13.5   The Residual Scatter

This section shows that the scatter about a fitted line can be quite complex. The x/y relationship within each set of readings does not have to be the same as that between the means $\bar{x}$, $\bar{y}$ of the different sets of data. But this does not affect the systematic $\bar{x}/\bar{y}$ relationship.

One Variable Controlled

This sub-section deals with the special case where one variable, x say, is deliberately controlled or selected.

(Statisticians are used to treating this as a special case of regression, as mentioned briefly in §12.7. But such data necessarily consist of different sets of data – one set for each selected value of x – and students find it easier when the topic is treated as a special case of BGA.)

When the controlled variable, x say, is subject to errors of measurement, least squares regression gives two numerically different answers for the regression of y on x, depending on whether one regresses the observed y's on the observed x's or on the "true" values of the x's . But in the BGA context nothing new or difficult arises. If x is controlled but subject to errors of measurement, Figure 13.11B shows that one is simply back to the general case discussed in this chapter, i.e. the relationship between $\bar{x}$ and $\bar{y}$.

13.6   Sample Data

There are no major problems with sample data in BGA. When the n's are small, the various means $\bar{x}$, $\bar{y}$, will deviate more from the fitted line.

The statistical significance of such deviations can be tested as outlined in this section. But even with large samples, one would not expect the $\bar{x}$, $\bar{y}$ to lie exactly on a straight line (or simple smooth curve). The problems of fitting the line (or estimating the parameters) are not radically different than with non-sample or population data .

13.7 Discussion

The BGA approach is not difficult for students who are new to statistics (and do not have to unlearn least squares ideas).

The bad news is that it usually takes many different sets of data to get anywhere much with BGA (whereas regression gives an instant "best" answer for any given set of data).

The general topic of lawlike relationships is discussed more in Chapter 19 and also in Part II of Data Reduction.

CHAPTER 14 : MANY VARIABLES

Students who may come across applications of multivariate techniques in their other reading or lectures need some acquaintance with at least the more popular forms (without aiming to turn them into expert analysts).

The techniques are probably more controversial than other aspects of our subject, and perhaps more complex. Depending on our personal views, we may want to warn against the techniques to a greater or lesser extent.

(In the Primer I may sound less critical of multivariate techniques than I feel. But teachers who want to emphasise the positive value of these procedures will, of course, be able to produce their own successful illustrations.)

14.1 Multiple Regression

This section assumes that we already have a small number of x-variables which we think are related to y. (Problems of selecting a fairly small set of x-variables from a larger number are touched on in §14.2.)

The first step taken in the Primer is that if the data are plotted on a graph, or arranged in order of y as in Table 14.2, one can usually see by eye which x-variables are related to y in any major way. (With small samples there can be questions of whether the correlations are significant.) The problem is how to describe the data by a single equation in y and all the x's together (if that is what one wants to do).

The numerical illustration used in the Primer is very small. If multiple regression is important in the students' subject-matter, one should familiarise them with a larger numerical example and also with what the resulting computer output can look like. For students in economics, management studies, etc., one will also need to touch on time-series and forecasting.

When one has fitted an equation like $y = 1.1 + 2.0x_1 + 1.3x_2$, this implies that the slope for x_1 is 2.0 whatever the value of x_2. Fitting the equation to sub-sets of the data with specific values of x_2 should therefore give parallel straight lines with slope 2.0 between y and x_1. But does the data show this? Is there enough data to judge?

Experience suggests that students can cope with discussions of such a question better than with the more purely "statistical" niceties or the mathematical formulae.

The Fit of the Equation

Much of the discussion in the literature concentrates on the overall fit ("What's the R^2?"). This is probably because people are often working with low r's between y and the individual x's and want to boost these. (Hence presumably also the urge to lump all the variables into a single equation.)

Given the traditional emphasis on R^2, it is worth reiterating how its numerical importance is often misinterpreted (just as for r^2 in Chapter 12). Thus for an $R = .8$, $R^2 = .64$ and this is said to mean that 64% of the observed variance of y has been "explained" or "accounted" for, which is supposed to sound fairly impressive. But a more descriptive measure of the fit of an equation is the size of the rsd relative to the initially observed s_y. The ratio of these two standard deviation is $\sqrt{(1-R^2)}$, which is .6. This means that even with an R as high as .8, the standard deviation of y has only been reduced by 40%. (Refer again to Table 12.2.)

14.2   Problems with Multiple Regression

The problems of bivariate regression recur with multiple regression (e.g. independent versus dependent variables and reliance on a "closest fit" criterion to determine slope-coefficients). How many "successful" multiple regression equations do we know?

Statistical authors rightly warn against interpreting the numerical coefficients in a regression equation as in any way reflecting the importance of the x-variables in question or their causal impact. (As always, correlation is not causation. The problem is worsened by collinearity.) Yet as noted in the Primer, everybody (including the statistical authors) then ignores these warnings ("If interest rates go up, inflation will come down ... "). Examples are needed to bring this home to students, preferably from their own subject-areas.

For example, suppose one has data on what people's weights are and what they eat and drink – the number of calories in their food, the amount of protein eaten, the amounts of specific vitamins and minerals (A, B, C, iron, potassium, phosphorus, etc), alcohol, tea, coffee, and so on. Will sticking such data into a multiple regression equation then show how much Vitamin B, say, affects weight? Does it matter whether the data analysed relate to different people (a cross-sectional analysis) or to changes in the same people (a longitudinal analysis)?

The Choice of Variables

The wide use of step-wise regression to select important variables highlights the highly "exploratory" nature of the approach. (If one had some prior knowledge of which variables mattered, one would not have continually to select variables in that rather mechanical way.)

One can say that some analysts appear to feel this to be useful but that it seems difficult to find many examples where it has led to results of lasting values, or where this has even been claimed.

One can suggest that students can use the following test in their reading: Has the reported result been successfully repeated in a wide range of different studies (as in normal science), or has it been obtained only once.

14.3   Factor Analysis

Many statisticians are highly critical of factor analysis and do not recommend using it although they seldom say explicitly what is wrong with it.

In principle there is nothing wrong with exploring how a set of variables interrelate. But in practice, there are much clearer ways of doing this, like noting correlational clusters. And if the patterns in the data are then not pretty self-evident, the results of using factor analysis to discover them are usually pretty dubious. In any case, once one has got to know one's subject-matter, how often does one still have to "explore" it by using factor analysis?

Factor Rotation

There are many ways of doing factor analysis (e.g. different rotations, and decisions about the number of factors to extract). The wide choice is nowadays less apparent because many people use Varimax.

Students are often struck by the arbitrariness and apparent triviality involved in "naming" the factors, especially if this is the final stage of the analysis.

Correlational Clusters

An alternative approach is to use right from the start the prior knowledge that is employed in factor naming, i.e. reordering the variables in the correlation matrix according to what we know or think about them and then seeing what clusters (if any) emerge.   This is illustrated in Table 14.11.

One should in any case go back to the correlation matrix after a factor analysis to show what it means in terms of the initial correlations. This will communicate better.   (If no clear correlational patterns were to emerge, it would throw doubt on the factor analysis.)   I know of many cases where factor analysts have not looked at their correlation matrices either before or after doing their analysis.

14.4   Other Multi-Variate Techniques

This chapter has gone further than many introductory texts in trying to familiarise students a little with two of the more popular multi-variate approaches.   It then mentions a few of the others only very briefly. Teachers may have to expand on this, using appropriate subject-matter examples.

14.5   Multivariable Laws

There are many successful scientific relationships in more than two variables.   They all stem from a slow building-up of knowledge over a sequence of different studies, rather than by trying to sort it all out in one go as is attempted in multiple regression or factor analysis.

14.6   Discussion

One feature that appears to be common to all methods of multi-variate statistical analysis is that although they are nowadays quite widely practised (especially with computers having taken much of the labour out of the analysis), there appear to be few results of lasting value. I personally know of none.

The explanation seems to me simple: The approach consists of applying a fairly complex analytic technique whose properties most users do not understand too well (and that includes me) to data which they also do not understand (otherwise they would not need an exploratory technique). The results are predictable.

Part Five: Communicating Data

The topics in this part of the Primer are seldom discussed in statistical texts.   But they are widely felt to be useful.   Problems of communication arise not only when we try to communicate data or results to others, but even when we are trying to understand data ourselves.

The topics – rounding, table layout, graphs and report writing – are not very technical.   There are no end-of-chapter glossaries and the chapters can be used as reading assignments if there is pressure of time in class.

CHAPTER 15 :   ROUNDING NUMBERS

The basic notion here is that we cannot easily manipulate numbers of more than two digits in our heads.   This is widely accepted.   But when digits are dropped there are trade-offs with accuracy.

People often are defensive about rounding: "I don't know what my data mean, but at least they are very accurate."   Successful analysis usually means deliberate over-simplification.   This also applies here.   The skill lies in knowing where to draw the line.

15.1   Rounding to Two Effective Digits

In my experience no one has claimed that they could mentally divide a three-digit number like 17.9% into 35.2%.   (One can quote the exception to prove the rule – a couple of mathematicians at a seminar at Purdue University Indiana who said they could.   But they gave different answers, so at least one of them was wrong.)

The distinction between "effective" and "significant" (i.e. non-zero) digits is important.   Using "effective" digits, defined as ones which vary in the given kind of data, copes with almost all the counter-arguments which people tend to raise.   The topic of variable rounding is a good example.

John Tukey has suggested recently a working-rule which copes with the special problems raised by our decimal system.   It is that the minimum variation shown in rounded figures should be about 1 part in 30. Otherwise one is over-rounding.

This covers the example of index numbers ranging from 117.9 to 144.3 in the Primer. Expressed as 118 and 144 means rounding to the nearest digit out of a range of 26, i.e. about one in thirty. That seems to be acceptable. (It is difficult to think of interpreting variations in the first decimal place like 117.9 or 144.3.) If the range had been substantially less than thirty, an extra digit should generally have been carried. If the range had been bigger, one could have rounded more, e.g. 118 to 444 is a range of 326 and expressed as 120 and 440 is still accurate to about 1 in 30.

15.2 Exceptions and Safeguards

Arguments about drastic rounding can be quite heated. The safeguards noted in §15.2 help to cool it (e.g. that the more precise data can usually be stored somewhere, just in case it is needed).

It helps to stress that there is a price to pay if one gives more than two effective digits – the data are less easy to take in and to use. (In other words, giving more than two digits does not deny that two digit numbers would be much easier.)

It also helps to confess that arguments over accuracy would almost disappear if one could generally give three effective digits. (More than three digits are virtually never needed.) But unfortunately, the human brain cannot cope with three digits.

Government statisticians often say they cannot round: "We produce a lot of data and have to do it very accurately, because we do not know what anybody will do with it." But no government statistician has yet spelled out a case where more accuracy than two (or three?) effective digits is needed.

(In a discussion at the Washington Statistical Society in 1978, one government statistician said that it was not necessarily their aim to communicate data clearly. The answer was that if they are actually trying to be obscure, they could surely do much better.)

15.3 Discussion

The idea of rounding is not new. In 1801 William Playfair wrote in his Statistical Breviary:

"As statistical results never can be made out with minute
accuracy, and that, if they were, it would add little to
their utility, from the changes that are perpetually taking
place, it has been thought proper in this work to omit the
customary ostentation of inserting what may be termed
fractional parts in calculating great numbers, as they only
confuse the mind, and are in themselves an absurdity. "

Following the rounding precept is greatly helped by knowing that
it is using more than two digits that "confuses the mind".   In the Harvard
Business Review of 1966, Robert Golde had not yet got this clear.   He
stressed the small loss in accuracy of simply "dropping off" digits, e.g.
that in reducing 21,742 to 21,000 the approximate loss in accuracy was only
3.41% !

(There was more discussion of rounding, and more examples, in
my 1975 Data Reduction.   But the phrase effective digits had not yet been
formulated then, nor the explanation of why our short-term memory cannot
cope with more than two digits.   This is set out more fully in "Rudiments
of Numeracy", Journal of the Royal Statistical Society A, 1977, 140,
277-297, and in "The Problem of Numeracy" The American Statistician,
1981, 35, 67-71.)

CHAPTER 16 : TABLES

The rules of table layout discussed in this chapter help make tables communicate better. Tables 14.7 and 14.11 in Chapter 14 already gave a dramatic example. (The main effect was rearranging rows and columns; rounding also helped.) The usefulness of row and column averages, as discussed in Chapter 3, can also be greatly stressed.

16.1 Ordering by Size

Ordering the rows or columns of a table by some measure of size often has the most dramatic effect of any of the rules. It can be laborious to do by hand, but is worth it. Recent computing developments will make it much easier, especially when interactive computing is linked with word processing facilities.

When looking at a table we usually do not know much about the numbers inside the table; we should order the rows and columns to help us with these. In contrast, we do not need to order the rows or columns to tell us some already well-known pattern in the row or column labels. (The alphabetical ordering in Table 16.1 is a classic example: We know that California comes after Alabama.) Objections can usually be met by saying that an alphabetical key ordering can help one find a specific entry (just like indices and tables of contents more generally).

Ordering by size not only helps us to see patterns and exceptions; it also provides us with instant correlations between the numbers in the table and the ordering variable, as illustrated by Table 16.3. It is a good idea to tell students how to look for a trend in such a table by forming sub-averages. For example, averaging the top five figures in the average column on the right of Table 16.3 gives $30/5 = 6.0$, the next five gives $34/5 = 6.8$. The answer is now easy to see: No simple systematic trend.

Ordering is a perceptual help because it gives structure. Thus Table 16.5A opposite, with the columns ordered by size, is easier to take in than the original Table 16.5. The patterns and exceptions stand out better, although ordering is not "essential", as noted on p.227 of the Primer. (The tables here are given in their original typed size for making an OHP transparency.) It would be interesting to see whether students now suggest that the two exceptional results for School C might simply be an inversion (72, 51 not 51, 72)!

TABLE 16.5   Rows and Columns Interchanged

		Bio-logy	Chem-istry	Eng-lish	Hist-ory	Math-ema-tics	Phy-sics
School A	%	49	18	73	45	23	34
" B	%	54	23	81	46	18	30
" C	%	72	21	51	53	25	36
" D	%	51	19	69	47	29	28
" E	%	48	16	76	52	21	42
Average	%	50*	19	75*	49	23	34

* Excluding School C

TABLE 16.5A   Columns Ordered by Size (and Row Averages)

		Eng-lish	Bio-logy	Hist-ory	Phy-sics	Math-ema-tics	Chem-istry	Ave-rage
School A	%	73	49	45	34	23	18	40
" B	%	81	54	46	30	18	23	42
" C	%	51	72	53	36	25	21	43
" D	%	69	51	47	28	29	19	40
" E	%	76	49	52	42	21	16	38
Average	%	75*	50*	49	34	23	19	40

* Excluding School C

The reason ordering by size seems to work so dramatically is again memory-factors. By having a simple order which we can remember easily (e.g. that population size decreases down each column), we can look at the detailed numbers in the table without having to strain our short-term memory so much.

In some cases there are practical difficulties or objections to ordering by size. Then the reason for not ordering has to be balanced against the loss in ease of perception.

16.2 Rows Versus Columns

It is widely accepted that reading down a column of single-spaced figures is easier than reading across. The effect is even greater with unrounded numbers. The reason again seems to lie largely in the short-term memory. Reading across, our perception is interrupted by irrelevant digits and blank spaces.

16.3 Table Layout

Once a table is typed we can see more clearly what the content is about and may well need to revise its lay-out. With word-processors, this will become physically much easier and hence more common.

The two main criteria are (i) to have numbers that are to be compared visually close together, and (ii) that details are easier to take in if the table has a general framework or structure (e.g. ordering by size and marginal averages to provide a visual and mental focus).

The point in the text about equally-spaced rows or columns in tables in general, and single spacing in particular (with deliberate gaps) is worth stressing. Short captions also help.

16.4 Verbal Summaries

Giving verbal summaries does not mean summarising the whole table. But some kind of "lead-in" to the table should be given, as a starting-point for the reader.

16.5 Discussion

References to "tables" in this chapter include informal working-tables, where the data are recorded for our own use. These are being increasingly computerised but often are not well laid-out.

When communicating to others (or having them communicate to us), we want "Information, not Data".   In other words, we need summaries rather than detail.

Few detailed tables usually need to be given.   If they are, they should be selective rather than comprehensive: to illustrate, explain, or allay doubts.   Much detail can therefore be relegated to appendices, filing cabinets, or data banks.

Tables which are directly referred to in the text (and which the reader is actually expected to look at) should not be bundled together at the back of the report as is sometimes done.   They should be given where they are needed.   Either they can be inserted in the text (preferably reduced 10% or 15% in size, which is easy and cheap with many photo-copying machines these days).   Or if there are <u>many</u> tables in a lengthy report, they can be given on the page facing the text, assuming that the pages can be backed when duplicating.

CHAPTER 17 : GRAPHS

Graphs are becoming increasingly popular, since computers make it easy and fun to produce them. But many graphs do not communicate the information they contain, or at least not easily. The chapter gives some criteria to distinguish between ones which work well and ones that do not.

To criticise some graphs does not mean that one is against graphs. I personally use plenty, e.g. in the Primer.

17.1   Showing Simple Results

The basic criteria are whether a graph communicates information to the reader, and whether it does so easily. It is widely accepted that graphs work well which have a qualitative story to tell – not detailed numbers – and a simple story-line. Such graphs can communicate their message in a dramatic and memorable way.

17.2   The Numerical Detail

A more controversial question is whether complex graphs can communicate. Even here it is pretty widely accepted that it is difficult to read numerical or quantitative information from a graph. However, this is not always acted upon.

Using some current examples from a newspaper (e.g. the New York or London Times), or ones from the course readings, students can be asked to try to summarise what the graphs are saying, in terms of both their main message and numerical detail.

It can also be useful to demonstrate briefly how difficult it is to work out any summary figures from such graphs, e.g. simple averages, deviations, etc. One would usually need to convert a graph back into a numerical table before any statistical analysis can take place.

17.3   Complex Story-Lines

It is often thought that the more complex a topic, the more a graph or picture will help to clarify it. In practice the reverse seems to be true. In any case, it is difficult to discover a structure or story-line from a picture if the presenter did not already know what this structure was and tried to communicate it.

17.4   Discussion

A question about complex graphs is why they are so popular both
with producers and users or readers. (People seldom complain about
graphs in the same way they complain about not liking numerical tables!)

Three reasons can be suggested to the class :

(i)   Graphs are popular with <u>producers</u> because one only has to
re-present the data, without any attempt at summary, analysis, or interpretation.   Very little intellectual effect is involved.   Furthermore,
computers can make it fun and physically quick and easy.

(ii)   Graphs are popular with <u>readers</u> because they feel they have
been given all the information (there it is on the graph), yet they do not have
to face up to any summary, analysis, or interpretation.

(iii)   More generally, many people do not realise how little
information they take in from a complex graph.   (As mentioned in § 17.4,
faced with a graph of some economic trends, we recognise well-known
features but nothing new is being communicated, at least not in a
memorable way.)

Complex graphs are difficult to take in because we have to absorb
a complex and unstructured set of information.   Thus in Figure 17.1, or
in Figures 17.3, 17.4, or 17.5, our eye has to move around a great deal,
to look at the axes, the captions, the shapes of the various curves or bars,
how high up they are, any numbers that are given, and so on.   In this
process we are generally putting an impossible burden on our short-term
memory.

We therefore have to remind ourselves by looking back a great deal
at things we have already looked at.   But complex graphs are so unstructured that we usually do not go back to the same starting-point.
This makes looking back confusing rather than reinforcing.

This problem of confused recall does not arise with <u>simple</u> graphs
which show a simple qualitative message – like Figure 17.2 or most of
the graphs in the other chapters of the Primer.

CHAPTER 18 : WORDS

Writing student papers, reports or memoranda and giving oral presentations are important skills for most of us. A little time can therefore be usefully devoted to this topic here, unless students have more extensive courses on it elsewhere.

Chapter 18 in the Primer concentrates on just a few major points. Since many statistics teachers do not have much teaching experience in this area, the Guide gives relatively more back-up information than for other chapters.

18.1 Starting At the End

Structure is most important. With technical writing the reader usually wants to know the findings before learning how they were obtained. He needs to be able to judge whether to read the paper at all, and if so, how much of it.

As readers we often have to turn first to the end to read the conclusions. This should not be necessary. The author knows what he has to say and should be able to write in the order in which he expects his piece to be read.

Usually technical writing should start not just with a summary (which is often too brief to communicate fully), but with the full description of the main results and conclusions (rather as in good journalism). If one then reads on, one has a mental framework on which to hang the details. This will also suit a mixed audience better (the less interested reader can stop early). Overall, it facilitates brevity.

However, writing in this way is not easy. We often have to think our problem through in a more historical sequence first (and possibly make an initial draft or outline accordingly) before we can "Start at the end" with our main findings and conclusions.

There are also exceptions: There may not be time to rewrite a paper in this order. (But that does not mean that it would not have been better for the reader if one had had the time.) Again, a textbook is different because it has a semi-captive audience. But even then it is best to have the major items near the beginning of each chapter. And whatever the structure, we need good sign-posting (like section and sub-section headings, formal or informal lists of contents, etc.)

18.2   Brevity

We all like brevity.

It is important to teach it.   In a written essay or appraisal of some
topic, students can be encouraged to make extensive notes first.   But they
should then argue through their views and conclusions on one side of one
piece of paper.   (I often enforce this myself;   in the end students appreciate
it.)   This is possible because students are writing for the teacher who
already knows the background and the problem.   They therefore do not have
to repeat it all.   The same situation holds true for most technical writing
(e.g. in business).

The main advice is: "When in doubt, leave out."   Every time.
Being brief does not make writing easier, but it should help the reader.
An old but effective quote is the end of a long letter (said to have been
written by G.B.Shaw or Mark Twain) "I'm sorry I could not make this
letter shorter, but I did not have the time."

Euclid achieved brevity by starting with his result.   He stated what
he was going to prove;   then he proved it;   and then he said "QED".
Instead of leading our reader at length through some argument or proof,
we can similarly start with our conclusion, e.g. that Product X in future
should be packed in metal cans rather than in glass bottles.   We then
follow this with our three main reasons: (1) ... ,   (2) ... ,   and
(3) ... .

All the other factors and analyses do not have to be discussed in
detail.   A brief listing will often do: "We also considered factors a, b, c,
and d, but they did not greatly affect the conclusions."   That indicates
thoroughness without parading all the work that has been done.   Having
done a lot of work, the writer does not have to tell everybody all about it.
Instead, one needs to consider what the reader actually needs to know, as
against what one feels one wants to say.   (If one wants a weighty report,
one can use appendices.)

18.3   Clear Writing

Verbose writing is often a sign of woolly thinking:

-   "It should be noted ..." usually means that the
    writer does not know why he is making the point.
    So "When in doubt, leave out", rather than bother
    the reader with an aside.

- "It is reasonable to suppose ..." always means
  that no reasons can be given (otherwise one would
  have said). One should therefore think again
  and rewrite or cut.

- Sentences which start with "Clearly, ..." are
  never clear (otherwise why say "Clearly ..."?).
  Usually they are not even true.

The Fog-Factor

The fog-factor outlined in Section 18.3 is a more general tool.
(It is very much easier to apply than other such measures in the literature.)
To calculate it for half a page or so of writing, I count the long words in
my head and the sentences on my fingers. After a time, writing with a
lower fog-factor becomes habitual.

The definition of the fog-factor is not water-tight. (E.g. are
there two syllables in "ratio" or three? What about names, numbers and
abbreviations?) But the precise definition of the index does not matter.
One is not playing games. One merely needs to ensure an average of not
more than two or three long words per sentence.

In technical writing we need long words (like statistics, regression
or correlation coefficient). But not too many. A fog-factor of two or
three still gives plenty of scope. It is tempting to make excuses ("My
topic is very complex ..."). But one can always reduce the number of
long words and cut up sentences into shorter ones. (Playfair's sentence
on page 70 is a marvellous construction; but after a page or two such
writing gets very heavy.)

Short sentences are good in their own right. The end of a sentence
tells us as readers that one can stop and think (something we often have to
do in technical reading). We may even want to read a sentence again.
This is easier if the sentence is short. A short paragraph similarly
allows us to go back and re-read several sentences without losing our
place on the page.

But not all sentences should be short. That would make for too
staccato a style. So variation is good: mostly short sentences, and a
few longer ones. But long sentences should have a reason, like giving a
qualification or illustration before the reader is allowed to stop and think.

18.4   More Than One Draft

Problems in writing are that the writer usually does not know
beforehand precisely what he or she wants to say or how to say it, and
that one is much too close to it.

Students must be prepared to revise, often drastically. The only
question is how much effort to make oneself and how much to leave to the
reader.

One also needs to get a distance between oneself and what one has
drafted.   Having done all the work, one is inclined to tell everybody all
about it.   There are several ways of getting a fresh view: When handing
the draft to somebody else to read, the scales often fall from one's eyes
(Is it too long, etc?).   Again, when discussing it with someone else people
often say "What I meant to say here was...".   They should write this
down: it is what they had meant to say, but didn't.   But the main clues are
(i) to allow time to elapse, and (ii) to get criticism from someone else,
and (iii) to act on this.

18.5   Oral and Visual Presentations

Good sign-posting is especially important with oral presentations.
The listener cannot go at his own pace, or skip and skim, or look at the
conclusions first and then start at the beginning, and so on.   We all
remember how comfortable it is to listen to a talk where the speaker tells
what he is going to say, then says it, and in the end brings it all together
for us again.

Students sometimes feel that the audience will get bored going over
the details if they have already had the main findings early on.   One
answer is that in an oral presentation, one can build up to the main findings
somewhat more slowly than in a written piece.   (One has a more or less
captive audience.)   More importantly, if the details are boring, we can
cut them and talk for 10 minutes, not 40.

Rehearsal is like rewriting.   Going over what one will say several
times in a fairly realistic situation makes all the difference to our clarity
and confidence.   This does not mean learning the script word for word,
but practising what one is going to say, in full and out loud.   Preferably
one should practise at least once in the room to be used (or something like
it), and with one or two friends sitting out front.

Speakers who sound most at ease and spontaneous are usually ones
who have rehearsed most.   (When asked why we do not apply all this to

our classroom lectures the only possible answer is that we know our
lectures could be much better.   Most college or university teachers are
untrained in communication.)

Visual Aids

Visual aids often become visual hindrances.   Rehearsal and try-outs
are crucial, especially in strange surroundings.   (The problem is less
acute in routine lecturing where we know the classroom and the like.)
Unreadable slides or transparencies are an insult to the audience.

18.6   Discussion

Most advice on good writing is too specific to have much impact,
like "Use active not passive verbs".   The guidelines summarised in the
Primer - structure, sign-posting, brevity, short words, revision,
rehearsal in oral presentation, and good visual aids - can have a much
more pervasive effect.

Part Six: Empirical Generalization

Notions like generalisation, explanation, and causality are not discussed much in statistical texts.   Yet they are of practical concern.

The material has been put at the end of the book because students can assimilate it better then.   (This also makes it easier to omit the part from classroom teaching if time presses, letting students read it on their own.)   But if there is time, it is good to develop the discussion further. This can be done with extended illustrations from the students' areas of specialisation and a guest-speaker from that area who has practical experience of interpretation, explanation, model-building, or the like.

CHAPTER 19 :   DESCRIPTION

The emphasis in this chapter is on describing more than one set of data.

19.1   Compared With What?

A politician complaining that his town only has 12 doctors is implying that this is low compared with other towns (e.g. relative to its population size).   We interpret observed facts by making comparisons.

Students appreciate being given a simple interpretative framework: "Compared With What?" often becomes a catch-phrase.

19.2   How Knowledge Builds Up

Empirical norms arise from finding that a particular result holds under a wide range of different conditions of observation.

If we apply a certain chemical to some plants but not to others and the former grow more, we have an isolated finding.   It is usual to find out whether it occurs again; then we have at least the possibility of a more general finding.   In contrast, if the result does not recur, there is no possibility of any simple generalisation.

But a first replication is still very limited – more are needed.
As stressed in the text, replication means carrying out studies which
explicitly differ.   (Students have often acquired the view that replication
implies completely identical repetitions.   They tend to be surprised and
interested to note that this would be both pointless and impossible.)

19.3   Lawlike-Relationships

Illustrations of law-like relationships from the students' area of
specialisation would be good to discuss here.   (Boyle's Law has been used
as an example in the Primer because it is fairly well-known to many
students, and .– by going back three hundred years – is closer to the
current situation in most social sciences.)

One simple example for students to discuss is the degree of
generalisation and the limitations of the height/weight relationship,
log w = .02h + .76 in § 13.4 (Table 13.3 in particular).

Laws are Approximate

It is good to emphasise the approximate nature of all scientific
laws (hence the less pompous "lawlike relationships").

A nice illustration (due to Professor Gerald Goodhardt) concerns
balls rolling down an inclined plane.   As we learned in high school physics,
Newtonian mechanics predicts rather accurately how long this takes.
But in fact they always roll down a bit more slowly than calculated because
the figures ignore things like air resistance and friction.   Indeed, without
friction the balls would not roll at all, but slide.   How wrong could
Newton be?   (Yet his theory is descriptively not bad – a deliberate and
vastly successful over-simplification!)

The Problems of Prediction and Extrapolation

Prediction is generally used in a different sense in the statistical
literature than in ordinary science   (e.g. in connection with regression
equations).   Statistical writers usually mean making an assertion about y
for a given value of x selected from the population initially analysed.

In contrast, prediction in practical science and technology usually
refers to data from a quite different population, e.g. results to be
observed next year, under different circumstances.   These matters are
usually not much discussed in our literature.

Problems of extrapolation arise particularly when dealing with cross-sectional and longitudinal studies, as noted in § 14.2 of this Guide. An example is also discussed in § 6.9 of Data Reduction.

19.4   Discussion

Scientific explanations and theories rest on descriptive generalisations. While the Primer repeats again and again that "correlation is not causation", we may also want to stress that there can be no causation without correlation.

CHAPTER 20 : EXPLANATION

"It's all very well in practice, but how does it work in theory?"

Apart from intellectual curiosity we want to understand how things work because findings which we understand will probably generalise more widely.   We can also cope better when those findings break down (e.g. that we do not have to understand how a motor car works, except when it doesn't).

20.1   Explanation and Theory

Explanations and theories are often not very deep things.   This can come as a surprise to students who have not yet thought about it.

Explanation operates mainly by connecting the finding in question to other things which are already known.   The link is seldom directly causal.   There are always "black boxes" or other deeper mechanisms involved, as illustrated by the fertiliser example in the text.

It would be good to develop illustrations from the students' own subject-areas.   For example, in physics saying that "ice floats because it is lighter than water" is by itself not very deep.   It does not say what actually happens.   Yet it very helpfully fits an isolated phenomenon – ice floating – to the wider range of findings that things float depending on the relation of their density to that of the liquid.

20.2   Assigning Causes

In everyday life as well as in science and technology we always think in terms of causal mechanisms.

For example, when the light goes out, we assume that somebody pressed the switch, or the bulb has gone, or a fuse has blown – probably in that order.   We check on these possible causes (usually without understanding anything deep about electricity) mainly because of past experience. If changing the bulb or checking/mending the fuse does not work, we think of some other causal mechanism, like a general power-cut or that we have been cut off for not paying our electricity bill.

In these cases we have a vast amount of general experience to help us assign the possible causes.   But often our knowledge is far less advanced, as in the example of Schools A and B in the text.

20.3   A Process of Elimination

Reaching causal conclusions is a slow and difficult process because the arguments are mostly negative.   Possible explanations are eliminated, rather than a particular factor being positively "proved".   That is why matching and standardisation cannot lead to conclusive results.

20.4   Critical Scrutiny

To judge whether a certain interpretation might be true – e.g. that salt is bad for blood pressure, or that promiscuity leads to infertility – we can ask ourselves

(i)   what sort of evidence we would need to believe it;

(ii)   whether such evidence (e.g. from a series of well-controlled studies lasting ten or twenty years) could in fact have been collected and at what cost;

(iii)   why nobody mentioned it (if there is such evidence);

(iv)   what sort of evidence probably does exist and what we probably think of it.

Topics of this kind lend themselves to class exercises and discussion.

20.5   Discussion

Causal explanations are difficult and wearisome to reach.   The ones we feel sure of in ordinary life are usually based on enormous amounts of accumulated evidence.   There is no harm in stressing the difficulties. It would be wrong to pretend that successful science and technological advances are easy.

CHAPTER 21 : OBSERVATION AND EXPERIMENTATION

Students of introductory courses do not design many studies (not without further training). But they need to understand the distinction between observational studies and experiments and the need for series of investigations.

(As statisticians, we must not let ourselves be dominated by our special randomised designs. In practice these make up only a small part of what scientists and technologists generally do.)

21.2 Observational Studies

It is common to demote observational studies. This is unrealistic and unhelpful. Much scientific and techological knowledge has been obtained through observational studies (with astronomy as an extreme example). Progress is made through a sequence of studies.

"Control groups" are widely used (e.g. smokers versus non-smokers, or people exposed to exhaust fumes in cities versus those living in the country). This relies on naturally occurring conditions, but otherwise resembles experimental control,with its limitations (§21.2 and 21.3).

Regression Towards the Mean

The selection of control groups and the like is especially open to creating bias. The particular example of selecting high and low scorers, i.e. regression towards the mean, is an important example for students to appreciate, including the escape route of selecting on a third variable.

21.2 Deliberate Experimentation

The great strengths of experimentation are that it makes it possible to observe phenomena which would not happen normally, and to test specific hypotheses suggested by previous results or theory.

The use of experimental control groups is however no more fool-proof than in observational studies. For example, the observed difference may be due to how the control group was selected rather than to the nominal difference in "treatment".

Students are often surprised (or even shocked) at the explanation of the state of affairs illustrated in Table 21.2.   (It is not called Simpson's paradox for nothing;   Journal of the Royal Statistical Society, B, **13**, 288-241 (1951).)

Table 21.2A below gives a more extreme case.   It starts with the same 75% and 45% recovery rates as in Table 21.2 for the groups with the new and traditional treatments.   But now both groups who are given the new treatment recover a little less well (82% and 10% for the new treatment, compared to 90% and 15% for the traditional treatment).   Yet because there are so many more patients with good prognoses who receive the new treatment, the overall percentage of recovery is still much higher for this treatment (75% versus 45%).

Table 21.2A   (Slightly) Lower Recovery Rates for the New Treatment

	No. of Patients	Number who Recovered	Number who Did Not Recover	Percentage who Recovered
New Treatment				
Good Prognosis	180	148	32	82.2%
Poor "	20	2	18	10.0%
All Patients	200	150	50	75.0%
Trad. Treatment				
Good Prognosis	80	72	8	90.0%
Poor "	120	18	102	15.0%
All Patients	200	90	110	45.0%

We can also point out that it does not follow from Table 21.2a that the new treatment really has a worse effect than the traditional one. There could be other Simpsons at work.   Or more simply, the new treatment might have made patients become dehydrated.   If only it had been accompanied with extra liquid, the recovery rate might have been better than for the traditional treatment.

"Correlation is not causation":   Decisive conclusions are difficult to reach.   Ruling out possible explanatory variables (e.g. by extensive replication under different conditions) depends on the scientist's knowledge of his or her subject (the possible factors which might have to be considered, past evidence in similar or related situations, and so on), hard work, and creative flair.

21.3   Randomised Experiments

Fisher's invention of the randomised experiment was a splendid breakthrough in allowing the experimenter to eliminate many of the extraneous factors when making experimental comparisons.

Factorial Designs

One may want to elaborate on the possibilities, like the use of "blocks" (to use the agricultural terminology) to reduce error variation, an extension of the "paired readings" approach in §10.5. Or one could discuss Latin Squares, showing how three factors can be compared at the cost of a two-factor design (but new at the cost of not being able to disentangle higher order interactions).

A feature of factorial experiments worth stressing to the class is that if one incorporates factors which are not expected to affect the results, the generalisation of the findings can be increased at almost no cost. For example, in the design of Table 21.3, the 400 patients could be selected as five groups of 80 from five different hospitals (or something equivalent).

A comparable randomised design could then be used for each hospital, with 20 patients in each of the four sub-groups. If the results for the five different hospitals turned out much the same (which is what one would expect), it would show that the findings generalise to more than one hospital. These might differ from each other in specified ways, like size, age, catchment area, etc. However, if the five hospitals unexpectedly gave different results, that should be learned sooner rather than later.

Limitations of the Randomised Experiment

By this time students may think that randomised experiments are the greatest thing since sliced bread. It may then come as a bit of shock to stress that the great majority of scientific results (including everything before the 1920's) were not – and still are not – obtained this way, for reasons mentioned in the main text. These are features worth illustrating with examples from the students' own subject areas. Nonetheless, where appropriate and possible, randomised experiments can provide a major short-cut.

21.4   A Sequence of Studies

One limitation of randomised designs is that they cannot usually be applied to sequences of studies. There are many factors (i.e. the variations in the conditions from one experiment to another) which cannot be controlled, let alone randomised.   Hence we need to stress once more the inescapable role of the observational approach.

21.5   Discussion

Scientific methods cannot be tied down to some simple foolproof recipes.   On the one hand, deliberate experimentation has been very fruitful.   On the other hand, we can never avoid the observational approach.

APPENDIX A: STATISTICAL TABLES

Even before computers and pocket calculators we did not need the detailed tables which are so often given in textbooks. Table A1 in the Primer gives values for the Normal distribution from 0 to 3.9 by steps of .1. Even this small table is, I think, more detailed than any of us need for almost any purpose. On those rare occasions where more detail is needed one can go to a library or computer.

Small tables of logarithms and square roots are given in the Primer despite pocket-calculators and the like, as they can help less numerate students come to grips with the concepts.

There are no tables of the F-distribution because the Analysis of Variance has hardly featured in the text.

APPENDIX B: ANSWERS TO EXERCISES

The end-of-chapter exercises in the Primer are of two kinds: Either to give practice in numerical calculations, or to provide a better feel for some of the topics discussed in the text. Thus the answers are sometimes relatively lengthy and argumentative.